POLYSACCHARIDE DISPERSIONS:

CHEMISTRY AND TECHNOLOGY IN FOOD

FOOD SCIENCE AND TECHNOLOGY

International Series

A complete list of the books in this series appears at the end of the volume.

Polysaccharide Dispersions: Chemistry and Technology in Food

Reginald H. Walter

Department of Food Science and Technology

Cornell University

Geneva, New York

ACADEMIC PRESS

San Diego London Boston

New York Sydney Tokyo Toronto

This book is printed on acid-free paper. ♾

Academic Press
a division of Harcourt Brace & Company
525 B Street, Suite 1900, San Diego, California 92101-4495, USA
http://www.apnet.com

Academic Press Limited
24-28 Oval Road, London NW1 7DX, UK
http://www.hbuk.co.uk/ap/

Library of Congress Card Catalog Number: 97-074420

International Stanrdard Book Number: 0-12-733865-9

PRINTED IN THE UNITED STATES OF AMERICA
97 98 99 00 01 02 MM 9 8 7 6 5 4 3 2 1

Contents

CHAPTER 2

The Polysaccharide–Water Interface

CHAPTER 3

State- and Path-Dependent Properties

CHAPTER 4

Concentration Regimes and Mathematical Modeling

CHAPTER 9

Saccharides in Fat Replacement

Preface

Polysaccharides are continuing subjects of heightened interest as feedstock, health and dietary adjuncts, environmentally harmless additives, and alternatives for industrial commodities of dwindling supply or prohibitive cost. Because the food industry is arguably the largest consumer of polysaccharides, this book discourses the technology and processing of polysaccharides as they apply to food. One objective is to cultivate a fuller understanding of their responses to extrinsic stimuli, akin to those of synthetic polymers, whose study over the past 50 years has generated reams of theory. In addition, explanations of the response of small molecules need to be extended to polysaccharides, because, as Schwartzberg and Hartel (1992) succinctly stated, the physical chemistry of macromolecules applied to foods has been neglected (in the United States) until recently.

In an oblique way, polysaccharides reemerged as a subject of advanced study when the importance of conformations and glass transitions in food-product development was recognized. Subsequently, applied texts on biological polymers were printed. By focusing on food polysaccharides, this author undertook an easy assignment, inasmuch as there was only one solvent, water, to consider. Moreover, Nature and the industrial supplier, whether inadvertently or deliberately, fix the inherent properties of these polymer molecules before they reach the processor and consumer.

This text should be equally comprehensible to the student, researcher, plant manager, and layman with merely a modest technical background: It argues a common phenomenology of polysaccharides, albeit under different combinations of stimuli: It is intended to be a portal to elementary comprehension of everyday principles that are borrowed overwhelmingly from synthetic polymer science, and not just another assemblage of polysaccharide facts. The prose is supported by empirical and theoretical equations whose origins were not always heretofore identified to food technologists.

The author expresses his gratitude and thanks to Professor Paul Okechukwu of the Federal Polytechnic Institute, Oko, Anambra State, Nigeria, and to Professor M. A. Rao of Cornell University for their expert review of the sections on rheology. The assistance of the following participants is also gratefully acknowledged: Professor Y. D. Hang of Cornell University for his review of the section on enzymes; Ms. Roberta Wertman of the Bryn Mawr class of 1999 and Mr. Steve Comella of the University of Rochester class of 1996 for their magnificent assistance while summer interns in this laboratory. Special thanks go to Ms. Elaine Gotham of the Cornell University graphic arts department for her splendid reproductions and to Mr. Joe Ogrodnick and staff for their superb photographic contributions.

Reginald H. Walter

Symbols and Abbreviations[†]

A	area
A_{sp}	specific surface area
A_2	second virial coefficient
a^0	area per adsorption site occupied by a single molecule
a_w	water activity
a_x	cross-sectional area of adsorbent
α	expansion factor
β	slope
C	capacitance
$C_{p,v}$	heat capacity
c	concentration
c_r	residual concentration
c_i	concentration of the ith component
c_m	molar concentration
c_n	surface saturation concentration
c_o	concentration of solvent (water)
c_u	unfilled-sites concentration
c_0	concentration at position or time 0
c_1	concentration at position or time 1
ς	amount diffusing
c^*	critical micelle concentration
CMC	carboxymethylcellulose
χ	Flory–Huggins interaction parameter
D	fractal dimensionality
$\mathrm{D_o}$	dielectric constant of water
Da	Daltons
D	diffusion coefficient (diffusivity)
DE	degree of esterification
DP	degree of polymerization
DS	degree of substitution
d	differential change
$\mathbf{d_o}$	density of solvent
$\mathbf{d}_i$	density of dispersion containing component i
$\mathbf{d_s}$	density of sphere

[†] Defined by context of use.

dn/dc_i	refractive index increment
∂	infinitesimal change
δ	angle in degrees
Δ	integral change
E	energy, enthalpy
E_a	apparent activation energy of viscous flow
e	base of natural logarithm (2.718)
ϵ	strain
$\dot{\epsilon}$	strain rate
η	coefficient of viscosity (viscosity)
η_a	apparent viscosity
η_E	elongational viscosity
η_o	solvent viscosity
η_i	viscosity of dispersion containing component i
η^*	complex viscosity
η_θ	intrinsic viscosity under theta conditions
F	force
f	function of
f'	segment factor
f''	degree of carboxylation
f_c	frictional coefficient
G	Gibbs free energy
$\mathbf{G}'$	storage modulus
$\mathbf{G}''$	loss modulus
$\mathbf{G}^*$	complex modulus
$\mathbf{G}$	modulus
$g(t)$	time correlation function
γ	shear
$\dot{\gamma}$	shear rate
$\mathbf{H}$	hydrophilicity
I	scattered light intensity
I_0	incident light intensity
I_ψ	scattered light intensity at angle ψ
i	component i
i_s	ionic strength
K_z	ionization constant
K	consistency
K_p	partition coefficient
k	constant (generic to a specific equation)
kJ	kilojoule
L	liter
l	length dimension
Λ	conductivity
λ	wavelength
λ_0	wavelength of incident light
M	monomolecular weight
$\overline{M}$	average molecular weight
$\overline{M}_n$	number-average molecular weight
$\overline{M}_v$	viscosity-average molecular weight
$\overline{M}_w$	weight-average molecular weight

m_i	mass of component i
μ_i	chemical potential of the ith component in solution
μ_o	solvent equilibrium chemical potential above a solution
μ_o^0	solvent equilibrium chemical potential above pure water (standard chemical potential)
N	Avogadro's number (6.02×10^{23} mol^{-1})
N	Newton
n_a	number of adsorption sites
n_i	number of moles of component i; refractive index of sol containing component i
n_o	number of moles of solvent; solvent refractive index
n_p	number of particles
$\bar{n}_i$	mole fraction of component i
$\bar{n}_o$	mole fraction of solvent
n_1	molar concentration in phase 1
n_2	molar concentration in phase 2
n	refractive index
nm	nanometer (m $\times 10^{-9}$); called millimicron (mμ) in non-SI terminology
ν	exponent
O	scattering wave vector
o	solvent (water)
Ω	electrophoretic mobility
ω	frequency
P	property of a polysaccharide
p	pressure
pI	isoelectric point
p_o	solvent equilibrium vapor pressure above a solution
p_o^0	solvent equilibrium vapor pressure above pure water (standard vapor pressure)
pH*	pH at 50% neutralization
ϕ_i	volume fraction of component i
ϕ_o	volume fraction of solvent
$\phi_{i,\text{gel}}$	volume fraction of solute in a gel
$\Pi(\psi)$	shape-dependent partical factor at angle ψ
pi	3.142
π	osmotic pressure
ψ	angle
$\boldsymbol{\Psi}$	angular velocity
Q	charge
R	gas constant (8.317×10^{-7} erg mol^{-1} K^{-1}; 0.08206 L atm K^{-1} mol^{-1}
R_g	radius of gyration
R_h	hydrodynamic radius
R_θ	radius of gyration under θ conditions
R_ψ	Rayleigh ratio at angle ψ
r	radius
$\bar{r}$	average radius
$\bar{r}_\theta$	average radius of a particle under θ conditions
$\sqrt{\bar{r}^2}$	root mean square end-to-end distance

$\sqrt{\bar{r}_\theta^2}$	unperturbed radius
S	entropy
S_v	sedimentation constant in Svedberg units (10^{-13} s)
s	second
σ	surface tension
σ_i	surface tension of component i
σ_o	surface tension of solvent (water)
$\sigma_{o,i}$	interfacial tension
Σ	sum of
t	time
t_c	correlation (decay) time
$t_{0.5}$	half-life
t^1	relaxation time
t_o	flow time of a solvent
t_i	flow time of a dispersion containing component i
T	absolute temperature (degrees kelvin)
T_c	cloud point (critical solution temperature)
T_g	glass transition temperature
T_{gel}	gelation temperature
T_{gz}	gelatinization temperature
T_m	melting temperature
$\boldsymbol{\tau}$	turbidity
τ	stress
τ_0	yield point
θ	the theta state; the theta condition (temperature and/or solvent)
V, v	volume
V_{el}	elution volume
V_i	volume of dispersion containing component i
V_0	initial volume; void volume
V_o	volume of solvent (water)
V_s	volume of solvent in a column's micropores
V_{sp}	specific volume
V_t	total volume
V_θ	volume under θ conditions
V_ω	final volume of dispersion
v_c	coil volume
v_e	partial specific volume of electrolyte
v_{ex}	excluded volume
v_f	free volume
v_i	partial specific volume
$\bar{v}_i$	partial molal volume of component i
v_0	void volume
$\bar{v}_o$	partial molal volume of solvent
v_m	molar volume (volume of 1 mol of i)
v_{sp}	specific volume
v_θ	coil volume under θ conditions
v_o	partial specific volume of solvent
w	weight

w_i	weight of a hydrocolloidal particle (i)
X	weight percentage of solute
x_n	DP function
x	distance
x_0	initial distance at 0 time ($t = 0$)
x_1	final distance ($t = 1$)
ξ	interaction potential
Z	dissymmetry ratio
ζ	zeta potential
$\sim$	asymptotically equal to
$\approx$	approximately equal to

CHAPTER 1

Origin and Characteristics of Polysaccharides

I. Introduction

Polysaccharides are a class of biopolymers constituted with simple sugar monomers. Those used in commerce and industry are isolates from terrestrial and marine plants or are principally the exogenous metabolites of some bacteria; many are modified by partial organic synthesis, and a few are the product of total biochemical synthesis. Isolates from the same species, but from different cultivars, are remarkably chemically uniform (Jones and Smith, 1949). These so-called natural gums and mucilages have teleological significance in plant metabolism and function; one primary responsibility attributed to many of them is winter and drought hardiness—a consequence of their water-binding characteristics. When extracted and purified, they are a major food item that is universally recognized as safe for human consumption. They are additionally an important industrial, scientific, and medical commodity. In petroleum recovery, mixtures of polysaccharides and sand are pumped into oil-well crevices to provide transport channels for oil and gas. In science, polysaccharides are crosslinked for improved mechanical strength and acid, heat, and shear resistance, for use as adsorbents and ion exchangers. Glycotechnology is a currently active area of pharmaceutical and medical research on oligosaccharides for drugs.

Cellulose, the most abundant polysaccharide, is the structural component of plant tissues; starch is the energy compound stored predominantly in seeds and tubers; glycogen is the animal counterpart of starch, but with shorter, more numerous branches. Cellulose and starch cohabit plant tissues with hemicellulose, protoplasm, lipid, and mineral matter in an organization interrupted by intercellular spaces that can amount to more than 50% of the total volume of some fruits and vegetables. A number of useful polysaccharides and their origins are listed in Table I.

TABLE I
Origin of Polysaccharides

Source	Polysaccharides
Terrestrial plants	Starch, cellulose, inulin guar, karaya, pectin
Marine plants	Agar, algin, carrageenan, furcellaran
Bacteria	Xanthan, gellan, curdlan, dextran, cellulon
Fungi	Pullulan
Derivatization	Carboxymethylcellulose (CMC), methylcellulose
Synthesis	Cyclodextrins, polydextrose
Animals	Glycogen, hyaluronic acid, chitin

In the food context, lettuce, apples, oranges, melons, cucumbers, etc., are uniquely succulent, turgid and crisp, because a small quantity of polysaccharide provides the coherence and mechanical strength necessary to embody 80–95% water. Similarly, *in vitro*, fruit jellies are made possible with 35–50% water.

Polysaccharides perform numerous other functions in food: for example, in nongluten bread, tenacious polysaccharide systems retain CO_2. Henderson (1988) ascribes the advantages of methylcellulose in the baking formula to a rise in surface tension at elevated temperatures, as a result of thermal gelation whereby the surrounding methylcellulose walls enclosing the gas-filled space are strengthened. The mechanical strength of gelatinized starch makes possible new structures from the extrusion-cooking of starchy influents. Dietetic beverages, sauces, etc., are often dilute dispersions in which a polysaccharide imparts "thickness" or "body" to water. Syrups made with partial starch hydrolyzates compete with sucrose as sweeteners.

In food technology circles, it is customary to hear about the excellent flavor-release qualities of polysaccharide gels and their attenuating effect on the flavor intensity of others. There are distinct flavor-release differences between polysaccharides (Malkki *et al.*, 1993) for which Baines and Morris (1989) offers the simple explanation that flavor molecules undergo a "restricted mixing" between the interior and the surface of a (polysaccharide) system—less so for random conformations than for gels, due to an increased resistance to flow (viscosity) in the latter. Explained differently, the rate of diffusion of flavor molecules, from the interior volume to the surface of a dispersion where they are sooner detected sensorially, accounts for the difference in flavor-release properties of polysaccharides. Gels hinder diffusion and sols facilitate it; consequently, nongelling polysaccharides are more efficient flavor releasers than are gelling polysaccharides. The same sensory intensity of flavoring and sweetening substances requires a higher concentration in a more viscous system than in a less viscous system (Morris, 1987). Unlike many monosaccharides of which they are composed, polysaccharides do not give the sensation of sweetness.

The texture described as mealiness in fruit and vegetable products is imputed to pectin dissolution and migration from the middle lamella and the consequent separation of cells that then act as discrete units. The slippery feeling of some starchy foods is due to amylose leached from microfibrillar bundles in spherulitic granules during the first stages of water transport of molecular starch from the granules under the influence of heat (gelatinization). Human sensory responses to polysaccharides are largely a reaction to geometrical (size, shape, volume, surface area), optical (color, gloss), thermal (specific heat capacity, conductivity, diffusivity), electrical (conductivity, dielectric effect), and mechanical (stress) effects (Szczesniak, 1983).

Polysaccharides are not always beneficial to humans: for example, low-molecular-weight carrageenans are possibly toxic (Engster and Abraham, 1976); in sugar manufacture, crystallization inefficiency and fouling of filters are problems caused by dextran.

II. Physical–Chemical State

Fruits and vegetables *in vivo* are nothing more than living macromolecular systems or biocolloidal dispersions. Outside the living tissue, they are multi-component dispersions and suspensions. It is these physical forms, influenced by one or more critical variables, that have many applications. The kind and scope of the polysaccharide response to stimuli *in vitro* depend on the polysaccharide's chemistry, the intensive properties engineered into it as a result of extraction, purification, and modification, and its interaction with the solvent surroundings.

Many polysaccharides are at the same time polyalcohols, polyacids, and polyesters, composed of a topologically linear, main sequence of connected monomers in one of two anomeric shapes (chair or boat), multiply configured, with varying amounts of methyl, acetyl, pyruvyl, etc., substituents, and occasional branching at regular and irregular intervals. The most stable ring form is equatorial (Thompson, 1992). The main sequence of monomers may consist of repeating di-, tri-, and oligosaccharide units, where each prefix denotes the number of simple sugars that constitute a unit. The dominant sugar determines its classification as a fructan, glucan, galactan, mannan, glycuronan, etc. A homoglycan contains a preponderance of one anhydrofuranosyl or anhydropyranosyl monomer; a heteroglycan has significant content of more than one monomer. Appendages to the fifth carbon of the pyranose ring lie outside the ring and consequently enjoy unlimited freedom of rotation. The two secondary and one primary hydroxyl groups on the monomer ring offer opportunities for chemical substitution, with the objective of achieving altered, sometimes exceptional, behavior depending on the degree of substitution (DS). In polysaccharides, complete substitution (DS = 3.0) is difficult to achieve. Derivatized polysaccharides are commonly ethers and esters.

As a result of pendant or ring ionizable groups, polysaccharides are largely divisible into two broad classes—ionic and neutral: starch and cellulose are typical of the neutral group. The natural ionic groups in polysaccharides are uronic, sulfuric, and phosphoric acid groups occurring as mixed salts of alkali and alkali earth metal ions (Na^+, Ca^{2+}, Mg^{2+}) or methylated to varying degrees of esterification (DE). The DE may be given as a ratio of esterified (mostly methylated) C-6 carboxyl groups and the average derivatization sites per monomer or as a percentage of the total number of carboxyl groups in the molecule. Polysaccharide polyanions in the presence of excess cations, including H_3O^+, simulate some properties of neutral polysaccharides.

Neutral and ionic polysaccharides are amphiphilic (amphipathic) molecules whose functions (e.g., surface tension, density, water affinity) are dictated by their constitutive nature (chemical composition, the arrangement of atoms and molecules), extensive properties (e.g., volume, weight, energy content, viscosity, concentration), and solvent surroundings. The chemical definition of "constitutive" differs from the engineering definition that relates stress and deformation in governing equations. Constitutive properties are intensive properties, i.e., they are invariant functions of mass, whereas extensive properties rely on and change with mass. Properties that depend on the number of discrete units into which a solute has been subdivided are colligative properties.

The distribution of substituents and branches in polysaccharides may be random or uniform. Distributions are isotactic when the substituents are all on the same side of the main axis, syndiotactic when they alternate on either side, and atactic when they are located at random. Regularity in the first two tacticities is conducive to crystallization (Sperling, 1986). Positional isomerism can lead to dissimilar properties, e.g., sulfate in carrageenans influencing their ability to gel (Anderson *et al.*, 1968; Guiseley *et al.*, 1980).

A. Molecular Weight and Degree of Polymerization

The molecular weight (M) of a polysaccharide is the gram mole or molar mass of 6.023×10^{23} molecules (Avogadro's number) that ideally are of a single size. Extracted, isolated, and purified polysaccharides in the same class, from the same sampling source, are seldom uniform in shape and size and are therefore preferably characterized by an average molecular weight $\overline{M}$.

M and $\overline{M}$ are experimental quantities that are arguably less than those of the parent polymer, *in vivo*, and vary with the method of measurement, most of which involve expensive apparati. Depending on the analytical technique, either quantity may be a weight-average ($\overline{M}_w$), number-average ($\overline{M}_n$), or viscosity-average ($\overline{M}_v$) property. Methodologically, $\overline{M}_w > \overline{M}_v > \overline{M}_n$, each capable of differing by a factor of $1-2 \times 10^2$ (Tanford, 1961; Walter and Matias, 1991). All the methodologies have their advantages and disadvantages.

The high number of monomers constituting M and $\overline{M}$ is referred to as the degree of polymerization (DP). Being predetermined by the manufacturer's methodologies, the average DP is constant but widely distributed from the mean, and is thus unsuitable without fractionation for most experiments involving purchased samples.

M and $\overline{M}$ are not related only to mass; they are equally a function of charge, solvent activity, and concentration. Some generalizations can nevertheless be made about the DP: the lowest DP stimulates the highest kinetic activity at the same weight concentration; low DP polysaccharides are more "soluble" and reactive than high DP polysaccharides; a high DP is necessary for dispersion viscosity measurably above that of water.

M and $\overline{M}$ are alternatively characterized by the radius of gyration (R_g), which is visualized as the radius of a thin circle transversely excised from an imaginary molecular cylinder, having a proximal end fixed at the center and a distal end traveling randomly along the circumference. The mass density is highest at the proximal end (Tanford, 1961). Random-walk theory indicates that the distal end will eventually maintain an equilibrium distance in the vicinity of the proximal end.

Unlike simple reducing sugars and oligomers, polysaccharides do not normally reduce Fehling's and Benedict's solutions—the historical reagents for identifying reducing activity (Cu^{2+} to Cu^{1+} after boiling for 5 s and left standing). The polysaccharides and higher DP oligosaccharides are ordinarily chemically inert, relative to their corresponding monomers and repeating units, because the high ratio of monomers to reducing end groups masks the carbonyl function. For example, assuming amylose $\overline{M} = 10^4$ Daltons (Da) and amylopectin $\overline{M} = 10^6$ Da, these starch fractions each contain one reducing end for every 55.6 and 5560 monomers, respectively (assuming exclusively 1,4-α linkages). A Da is the unit of molecular weight equivalent to one-sixteenth the gram atomic weight of oxygen.

The importance of $\overline{M}$ is unmistakable when it is recalled that 180 g glucose in 10^3 g water [1 molal (m)] lowers the freezing point by 1.86°C, whereas the same weight of polysaccharide (1.80×10^{-3} m, assuming $M = 10^5$ g) has no such effect and, moreover, it is difficult if not impossible to disperse 10^5 g of any polysaccharide in 10^3 g water. Identical arguments hold for boiling-point elevation and osmotic pressure. However, polysaccharides at much lower concentrations exercise influences, most prominently by structuring water into high-viscosity fluids and obstructing the crystal order at and below 0°C.

B. Configurations and Conformations

The earliest studies of natural polymers, predating those of synthetic polymers, elaborated an association hypothesis of a primary macromolecular structure held together by physical bonds (Purves, 1943). It was later proven that the primary bonding is covalent and that a tertiary structure results from physical bonding between primary structures that are themselves high-molecular-weight compounds. Subsequently, "well-defined generalizations" emerged as the foundation of a new discipline—polymer science (Flory, 1953). Advances in synthetic polymer research have outpaced those in biopolymer research.

The primary structure of a polysaccharide is the main sequence of connecting sugar monomers covalently linked as α and β glycosides. The constitutively fixed bond lengths and angles controlling the ring orientations comprise a secondary structure (configuration) whose effects are most acute in heterobiopolymers like pectin and carrageenan, for example, wherein rhamnose in the "hairy" regions of the former (Thibault *et al.*, 1991) and sulfate in the latter (Stanley, 1990) cause kinking. Interchanges of atoms in the primary and secondary structures can occur only by rupture and reformation of valence bonds. The single-bondedness of the glycoside linkages enables different segments to rotate independently in solid, liquid, and gaseous space in the manner of a freely jointed chain, unlike those of synthetic polymers where there may be bond and steric hindrances. Independent segment rotations culminate in a tertiary structure (conformation). Polysaccharide quaternary structures (Fishman *et al.*, 1991; Dea *et al.*, 1972) consist of clusters of tertiary structures held together by a net sum of intermolecular forces (hydrophilic, electrostatic, ionic, and hydrophobic) under various influences of solute concentration, solvent, electrolytes, temperature, and shear. Tertiary and quaternary structures are physically reversible. The DP exerts much influence on conformation. A low DP is conducive to the development of rodlike geometries and crystals, whereas a high DP is conducive to random-coil and amorphous behavior. The highest incidence of coiling is found in the amorphous regions in linear polymers (Sperling, 1986).

Hydrophilic, electrostatic, and ionic forces are dominant in polysaccharide systems, given their constitution; less dominant is the hydrophobic force that is more likely to be encountered in proteins and other biochemical multisubunit particles (Ben-Naim, 1980).

Linearity of the polysaccharide primary chains facilitates parallel ordering of the tertiary structures in two (e.g., β sheets of cellulose) and three (e.g., helices in xanthan gum) dimensions that grow in time to floccules.[1] The growth of these supramolecular associations (Finklemann and Jahns, 1989) is accelerated or retarded by environmental stimuli. Floccules—not single molecules—dominate macromolecular activity above a critical solute concentration. In the words of Doi and Edwards (1986), "the macroscopic properties (of a polymer liquid) depend only on a few parameters specifying the molecular characteristics, and insofar as these parameters are the same, different systems behave in the same way."

Parallelism in localized regions of polysaccharides creates three-dimensional order between segments of the primary chain that is conducive to crystallite formation, which in turn has a profound effect on the polysaccharide response to ambient stimuli. A high incidence of crystallinity spawns refractoriness; a low incidence contributes to amorphism. X-ray diffraction of cellulose has shown that oxygen–oxygen distances in one direction of the tertiary structure (the unit cell) are shorter than the van der Waals diameter of oxygen atoms in an organic micromolecule. This suggests strong lateral forces (Tanford, 1961) operating at these sites, accounting for refractoriness. A now virtually discarded fringe micelle model describes polymer crystals as consisting of ordered regions bonding in series with amorphous regions in the same molecule (Cowie, 1991). Contemporary theory favors crystallite formation from single molecules acting as crosslinks between other linear molecules (Severs, 1962). Natural and artificial events (e.g., branching, kinking, bulky substitution, high cosolute additions) disrupt the tendency toward parallelism, crystallite formation, and refractoriness.

Supramolecular assemblies assume smectic, cholesteric or nematic orientations, whereby the planar axis may be perpendicular (smectic) or parallel (cholesteric) to the molecular axis. In a nematic orientation, the assemblies are not planar, although the molecular axes are parallel (Elias, 1979).

Parallelism and stereoregularity are in harmony with the kind of molecular order that ends in helix as well as crystallite formation (Billmeyer, 1984; Cowie, 1991; Rinaudo, 1992) in amylose (1,4-α), cellulose (1,4-β), and curdlan (1,3-β), for example—three stereoregular polysaccharides that do seem to favor these conformations. Deesterified carboxyl sequences in the primary structure of low-methoxyl pectin and between poly-L-guluronate sequences in alginate facilitate bridging by Ca^{2+}, illustrated in the "egg-box" model of gelation (Rees *et al.*, 1982). Rinaudo (1988) interprets the effect of this blockwise deesterification on pectin gelation to mean a requirement of a critical, minimum length of a continuous series of carboxyl groups. Rhamnogalacturonan blocks are not organized into the conformation prerequisite to gelation (De Vries *et al.*, 1981), because kinking (Rees and Wright, 1971; Oakenfull, 1991) precludes the parallel orientation. Guar and locust bean gums are gelling and nongelling galactomannans, respectively; the former

1. Flocculation, coalescence, coagulation, and aggregation are herein used synonymously.

has more extensive, uniformly spaced substituents. In cold water, guar gum hydrates more than locust bean gum, in an apparent contradiction of the expectation of regularity in the fine structure. As exemplified by the hydratability of these two gums, polysaccharide properties are integrals of many factors.

Polymers are modeled on three idealized conformations, viz., a solid sphere (Harding *et al.*, 1991a), a random coil, and a stiff rod (Doi and Edwards, 1986). Linear polymers are more or less randomly coiled in solution (Eisenberg and King, 1977); so are polysaccharides (Morris, 1976), inasmuch as the freely jointed chain offers them endless possibilities for rotation averaging conformations within the extreme bounds of a flexible coil and a rigid rod. Computer modeling suggests an improbability of very short and very long distances between proximal and distal groups; most fall instead at intermediate distances from each other. The almost infinite number of possible segment associations must be treated statistically (Smith, 1982). It is the thermodynamics of the particular system that dictates whether polymer molecules acquire the habit of a chain or rod (Rinaudo, 1988). When there is interaction, the resulting conformation may not necessarily acquire a state of minimum energy (Tvaroska *et al.*, 1992). The study of Tvaroska *et al.* (1992) conveyed the picture of paired molecules (of κ-carrageenan and mannan), approaching each other in their respective ground states, changing conformation, and maximizing the van der Waals and Coulombic attractions.

Solvent conditions have a profound influence on conformation, as proven by glutamic acid, a random coil at high pH, but a helix in acidic media where ionization is depressed (Doty *et al.*, 1957), and by hyaluronic acid, a random coil in dilute solution at neutral pH but very stiff in acid (Shah and Barnett, 1992). A perfect helix can exist as a random coil at a sufficiently high DP (Elias, 1979). According to Alfrey *et al.* (1942), a solvent manifests its effects through an unbiased, mean statistical configuration that shifts between a completely curled (in an energetically unfavorable solvent) and a completely extended (energetically favorable solvent) geometry at constant temperature. An energetically unfavorable solvent is described as "poor," and an energetically favorable solvent as "good." A good polysaccharide solvent maximizes and retains an expanded conformation, as a result of solute–solvent interactions, at the expense of solute–solute interactions: a poor solvent shrinks the molecular volume and retains the polysaccharide as a compact coil.

Water, occasionally modified by a small volume of ethanol and electrolytes, is the universal food solvent. In this medium, a polysaccharide polyanion's molar volume (v_m) is maximum, because charge repulsion forces the molecules into the stiff (uncurled) conformation of a rigid rod that cannot be extended further, but can be oriented (Odell, 1989). v_m is larger for extended, stiff conformations than for compact random-coil conformations (Berth, 1992). As solvent quality deteriorates, e.g., by addition of salt or ethanol, molecular motion declines, v_m becomes less, and the solute molecules begin to exercise a strong attraction for each other. The condition determined by temperature and solvent that allows neither polymer–polymer nor polymer–solvent attraction is the theta (θ) temperature and θ solvent, respectively. A nonsolvent is any organic, water-miscible liquid (e.g., ethanol and acetone) that inclines a dispersed solute toward the θ state.

The rigidity (or flexibility) of dispersed polyanions is basically a function of the solvent's ionic strength through its influence on electrostatic repulsion (Eisenberg and King, 1977). Hence, a polyanion is most rigid in pure water where, at any instant, the fixed negative charges foster an equilibrium segmental and counterion distance at its farthest. Excess cations relax the rigidity by shielding the charged sites, thereby allowing the primary chain to revert to a flexible coil (Miles *et al.*, 1983). Neutral polysaccharides experience no charge restriction and therefore freely migrate toward and away from each other in dilute solution.

Stiffness can cause polysaccharide polyanions to appear as pseudocrystals (Barnes *et al.*, 1989), because they easily order themselves into an apparently crystalline solid at rest and revert to a liquid when agitated. The 1,4-β-glycoside linkage is inherently stiff and extended (Blackwell, 1982): such molecules incur anisotropy.

For a constant polysaccharide mass, an extended (random) coil exposes more surface area than does a helix, and a single helix exposes more than a double helix. The energy content of a polymer molecule is a property of its surface area. Thus, one consequence of a coil-to-helix transition is a diminution of the macromolecular exposed surface area and energy in compliance with the law of entropy. An increase in viscosity coincides with an increase in surface, inasmuch as the resistance to motion covers a wider area.

1. Specific Conformations

Depending on the DP, charge, and solvent conditions, CMC may be a random coil or rod (Guo and Gray, 1991). According to Mitchell and Blanshard (1979), short chains in alginate and pectate gels are stiff and extended, while those of higher DP are flexible. Other specific conformational descriptions of dispersed polysaccharides are random coils for konjac mannan (Kishida *et al.*, 1978) and guar gum (Robinson *et al.*, 1982), rods for gum ghatti (Srivastava and Rai, 1963; Elworthy and George, 1963), stiff coillike for starch (Banks and Greenwood, 1975), ribbon (Ring, 1982; White, 1982), stiff extended ribbonlike, and planar for cellulose (Glass, 1986) and chitin (Blackwell, 1982), extended ribbon for alginic acids (Atkins *et al.*, 1970) and cellulose (Blackwell, 1982; Belitz and Grosch, 1987), stiff coil for agar (Hickson and Polson, 1968), short stiff spirals for gum arabic (Meer, 1980b), single helix to triple helix for curdlan (Marchessault and Deslandes, 1979; Stipanovic and Giammatteo, 1989; Ross-Murphy, 1991), double helix for carrageenans and agarose (Ross-Murphy, 1991), and left-handed double helix for agarose (Arnott *et al.*, 1974). Veis and Eggenberger (1954) suggested the possibility that branching was responsible for the stiff arabic acid coil. The agarose gel model of Hayashi and Kanzaki (1987) contained alternating helix and kink segments. Some polysaccharide molecules, e.g., agarose (Hayashi and Kanzaki, 1987), hardly change dimension.

Pectin conformation may be a strongly pleated ribbon (Belitz and Grosch, 1987) or a random coil (Axelos *et al.*, 1987; Berth, 1988; Berth *et al.*, 1990), but the data of Harding *et al.* (1991b) on a citrus pectin (approximately 70% anhydrogalacturonic acid in 0.03-M phosphate buffer, $\overline{M} = 2.0 \times 10^4 – 2.0 \times 10^5$) fitted the model of a rod with some anomalous features indicative of a wormlike coil. The wormlike conformation is described as different orders of flexibility, from a random-walk chain to a stiff rod (Dautzenberg *et al.*, 1994). Contemporaneously, Hourdet and Muller (1991) indicated that a flax pectin (80% anhydrogalacturonic acid in 0.2-M sodium chloride) in an inclusive molecule-weight distribution range behaved like an extended coil. Pectin random coils are further described as continuously mobile (Morris *et al.*, 1981).

Amylose has been described as a double helix (Gidley, 1989) and a random coil (Ring and Whittam, 1991). The prevailing hypothesis is that it is a flexible random coil of extended helical segments connected by deformed nonhelical segments (Szejtli, 1991). Gidley (1988) identified two molecular structures in an amylose gel—one of relatively immobile double helices and the other of all energetically permissible conformations. Cellulose exists in helical conformation in the crystalline state (Guo and Gray, 1994).

Double helices have been implicated in the gelation of many polysaccharides, e.g., agar, carrageenan (Rees, 1972a, b; Glass, 1986), and gellan (Chandrasekaran *et al.*, 1988a). This last gum is proposed to exist in solution, under nongelling conditions, as a disordered coil at high temperatures and as a double helix at low temperatures (Robinson *et al.*, 1991). Marine polysaccharides are often suggested to be in a double-helix conformation (A. H. Clark, 1992). Two forms of carrageenan (κ and ι) are random coils above 50°C and double helices below (Anderson *et al.*, 1969); these are the structural units that must then aggregate preceding gelation (Morris *et al.*, 1980). The gelling mechanism of alginate has been deduced as an association of double helices made up of guluronate moieties that bind divalent ions cooperatively (Rees, 1969; Sime, 1990). Callet *et al.* (1988), unable to detect any disassembly of a xanthan double helix, explained xanthan rheological properties on the basis of preexisting, ordered conformations without change in molecular weight.

Curdlan adopts a triple helix conformation in the dispersed and solid states (Deslandes *et al.*, 1980) and reverts to a random coil in 0.25-M sodium hydroxide (Stipanovic and Giammatteo, 1989). Some glucans are ordered in dilute alkali and disordered at higher concentrations (Ogawa *et al.*, 1972).

Braudo (1992) showed that helix formation is not a prerequisite to gelation: rather, gelation may result from hydrogen bonding and from intermolecular coordination binding of cations (e.g., calcium in the case of low-methoxyl pectin and potassium in the case of κ-carrageenan): for example, xanthan is a random coil at high temperatures and a helix at low temperatures (Griffiths and Kennedy, 1988); upon heating, its dispersions undergo a change of state from an ordered helical (Rees, 1972a, b) to a disordered coil structure. The conformation is influenced by the degree of ionization—helical below 0.85 and random coil above (Young and Torres, 1989). In the native state, this gum has also been described as rigid and rodlike (Sandford, 1979) in single, fivefold helix conformation (Griffiths and Kennedy, 1988) that converts to a random coil at high temperatures in low-ionic-strength media (Morris *et al.*, 1977). Upon cooling, the molecules revert to their native conformation and align themselves with unsubstituted regions in galactomannans (Dea and Morris, 1977) in a gelling mechanism of cooperative crosslinking.

Xanthan gum was shown to be stiffer than CMC and alginate: all three are ionic polysaccharides, with CMC having slightly more flexibility than alginate under identical conditions (R. C. Clark, 1992). The invariant nature of xanthan dispersion properties is attributed to the stability of the tertiary structure. The indifference of this gum to salt is explained by its already rigid conformation (Morris, 1976).

Concentration has a role in conformation: dextrins are random coils in dilute dispersions and compact coils in less dilute dispersions (McCurdy *et al.*, 1994).

2. *Interaction, Disorder–Order and Order–Disorder Transitions*

In the absence of a formal definition, "interaction" is understood to involve reciprocal action between two polymer molecules: in the process, one or both undergo a transition from an ordered to a disordered state or vice versa. Gelatinization (Biliarderis, 1992; Kokini *et al.*, 1992) is an order–disorder transition of starch granules interacting with water in a critical temperature range. The initial, intact spherulites have the organization of crystalline regions of relatively short, hexagonal fibrils surrounded by amorphous regions. In the first stage of the transition, the granules hydrate and swell; in the terminal stage, they lose rigidity, become disordered, and disintegrate. Finally, fluid starch exudes from the physical confines of the former granule into the outer volume of water. Complete gelatinization is irreversible. After a period of time (aging), the dispersed (disordered) amylose reverts to a differently ordered association (retrograded starch), whereby linear molecules and segments are returned to some semblance of an organized crystallite, albeit vastly dissimilar to the initial structure. Retrogradation is irreversible in ordinary storage, but can be reversed by heat and moisture intervention.

The interaction between xanthan and konjac mannan gums in an aqueous sol leads to gels in an ordered conformation at 42°C and in a disordered conformation at 57°C (Williams *et al.*, 1991). Thermoplastic gels are a reversible disorder–order transition and thermosetting gels are opposite and irreversible. The significance of all these conformational transitions is the dependent change in properties and function concurrent with a change of phase, location, size, and/or energy status: the final phase, location, size, and/or energy status is not identical to the initial.

The order–disorder transition of a polysaccharide crystallite is loosely referred to as melting, notwithstanding the absence of the solid–liquid phase change, as in the melting of fat for which the word was reserved. Broken-curve heating (Hersom and Hulland, 1981) in soup and juice concentrates, whole-grain corn, starchy fluids, etc., results from an order–disorder transition whereby a solid or gellike mass is first heated by conduction and then by convection as the energy content and subsequent mobility of the mass increase.

C. Colloidal Activity

By definition, a colloid has at least one molecular or macromolecular dimension in the 1–500-nm range (Jirgensons and Straumanis, 1962). Polysaccharides fit this definition. The nominal width is the orbital diameter in atomic proportion to the hydrated molecule rotating around its long axis. Heyn (1966) reported widths of 25–40 Å (2.5–4.0 nm) for single cellulose microfibrils; Kanzawa *et al.* (1989) reported 50–250 nm for gelling polysaccharides and 10–20 nm for nongelling polysaccharides. Pfannemüller and Bauer-Carnap (1977) measured an average of 10 nm (100 Å) for fibrils and fibrillar aggregates of amylose, DP = 100–7200. According to Harada *et al.* (1991), microfibrils in curdlan gels comprising subunits 2–3 nm wide are themselves 20–25 nm wide. Crystallites approximate a length of 46 nm and a diameter of 7.3 nm (Cowie, 1991). Cellulose microfibrils can be $10–10^3$ nm long (Weibel, 1994). Fishman *et al.* (1986) determined length-to-width ratios of 120–200 per pectin.

Polymers, organic and inorganic molecular clusters are synonymously called macromolecules to distinguish them from the smaller sizes of organic acids, salts, ethanol, etc. (micromolecules). Properties unique to organic macromolecules begin to be manifest at $M = 10^3–10^6$ Da (Billmeyer, 1984; Sperling, 1986).

1. Dispersibility

Polysaccharides are not soluble in water in the classical sense that micromolecules (e.g., sodium chloride and sucrose) are. In micromolecular solutions, solute and solvent are indistinguishable to the solubility limit, and

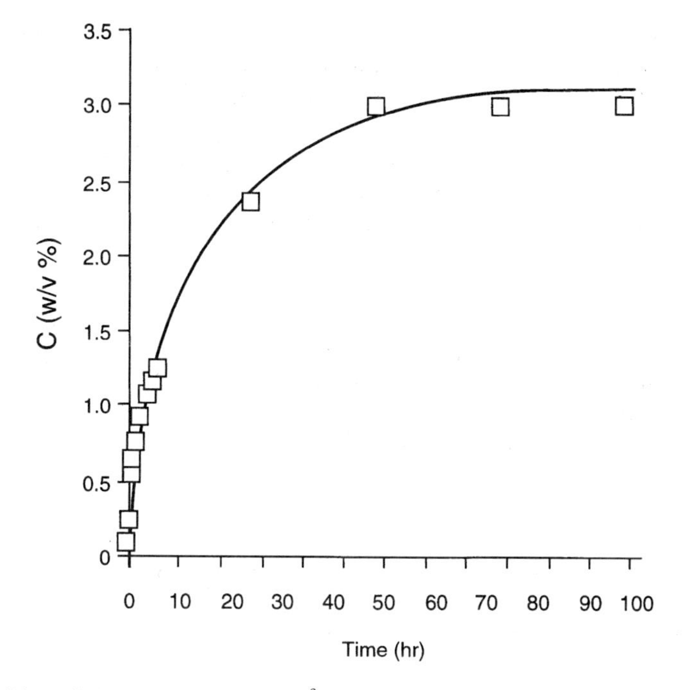

Figure 1 Dispersibility of 50 g pectin in 10^3 mL water in a 2-L two-neck distilling flask held at 26°C with mild stirring. The weight per volume percentage concentration was calculated, after periodically withdrawing 10-mL samples from the suspension and oven-drying them to constant weight.

the solutions are described as homogeneous, because the two components merge into a monophase. Polysaccharide–water systems in any solute–solvent ratio do not have a solubility limit; instead, they are heterogeneous dispersions called sols, possessing an interface, albeit imperceptible at times. Polysaccharide sols may, however, exhibit a pseudosolubility limit where, for every molecule diffusing out of the solvent phase, one enters it, until the system is disturbed and more solute leaves the liquid phase than enters it, or solute begins to be deposited on existing dispersed solute to the point of phase inversion (i.e., a solid-in-liquid to a liquid-in-solid transition). In the latter, the rate of diffusion is initially fast and then slows as the pseudosolubility limit is approached (Fig. 1). From the metastable pseudosolubility limit, precipitation is possible, given the appropriate stimulus.

2. *Hydrophilicity*

Polysaccharides, remarkable for their water affinity, can embody many times their weight or volume of this solvent—hence the name hydrocolloids. Polysaccharide dispersions are either hydrosols, hydrogels, or xerogels (dehydrated hydrogels). The high rate and capacity of absorption mimic swelling that is characterized by a swelling ratio defined as the quotient of the volume of dispersion containing component i (V_i) at maximum hydration and the initial volume (V_0) of a xerogel. The absorbed water is customarily referred to as water of hydration: it has a lubricating and plasticizing effect on human ingestion. Bulkiness and the feeling of satiety in the human regimen, derived from consumption of fibrous fruits and vegetables, are sensory reactions to a large V_i/V_0. Hydrated tissues are more digestible than dehydrated tissues, because of the softened texture resulting from water of hydration.

Polysaccharide polyanions swell to many times their volume at θ (V_θ), more than do neutral polysaccharides, and shrink proportionately: V_i/V_θ is markedly enhanced by water and depressed by nonsolvents and cations through their action on V_i.

Moisture is lost linearly from dehydrating plant tissues to 20–30% (Fig. 2), indicating that some water of hydration is loosely held. Apparently

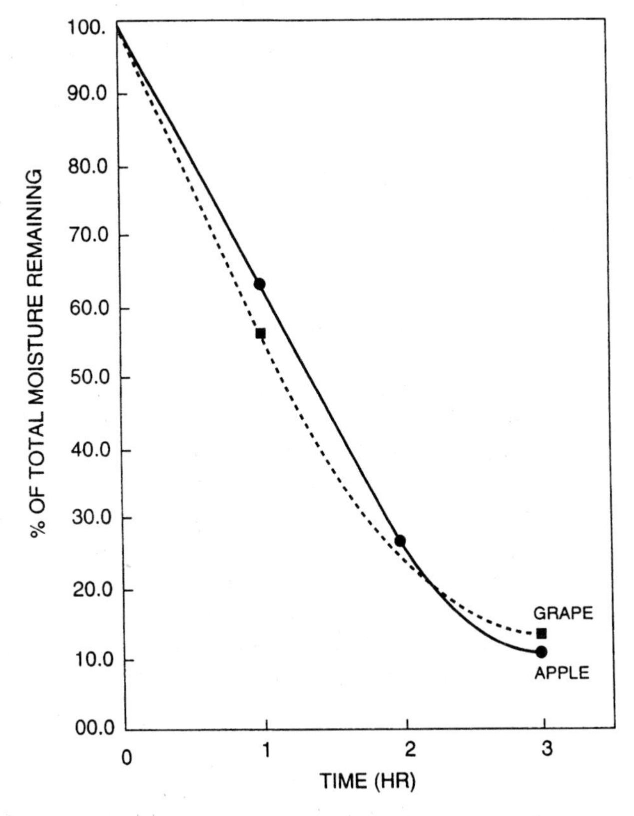

Figure 2 Drying curves for grape and apple pomace held at 105°C. The percentage moisture remaining was calculated, after withdrawals of 10-g samples from a 10^3-g batch and drying them at the designated intervals.

dehydrated tissues can retain single-digit moisture levels, while appearing to be completely dry. This property is exploited in the use of polysaccharide xerogels as humectants, e.g., anhydrous starch added to baking powder to retard caking.

3. Surface Area and Surface Tension

A dispersed hydrocolloid exposes an inordinately large surface area relative to its volume or weight. The relationship between energy (E) and the total exposed surface (A) is stated as

$$E = \sigma A. \tag{1.1}$$

σ is the surface tension of the hydrocolloid in an energy per area unit. As a result of σ, the surface molecules are attracted toward the interior of a substance: the larger the value of σ, the stronger is the internal cohesion and the less is the surface area. A dispersed polysaccharide spontaneously seeks to lessen A, in compliance with the second law of thermodynamics pertaining to energy minimization. It follows from Eq. (1.1) that fine subdivision of a polysaccharide solute requires energy: the corollary is that the more surface a dispersed polysaccharide exposes, the more energetic it is and, consequently, the less stable it should be. The interfacial tension ($\sigma_{o,i}$) is the equivalent of σ when the substance interfaces a substance other than air.

4. Viscosity, Elasticity, and Viscoelasticity

The resistance that macromolecules encounter as they flow past each other in a solvent is called viscosity: the larger the surface area, the higher is the viscosity. Random coils above θ, exposing a larger surface area than other configurations, give a higher viscosity; compact random coils offer less resistance to flow (lower viscosity) than extended rods. The energy expended during flow, less at high temperatures (Severs, 1962), is not recoverable. The viscosity created by infinitely thin parallel fluid planes transported co- or countercurrently is called simple shear viscosity. Planar flow may be initiated in other geometries, e.g., rotational, telescopic, and twisting (Van Wazer *et al.*, 1963). Humans evaluate simple shear viscosity as the sensation of thickness. Certain fluids, e.g., honey and paint, are quite resistant to flow and are therefore considered to be very thick or viscous. Most polysaccharide dispersions increase in simple shear viscosity with the DP and concentration and decrease with increasing temperature.

Viscous transport is laminar or streamline, if molecules in the (imaginary) planes of fluid in any geometry, flowing along a velocity gradient, do so with the same translational and rotational velocities. Streamlines are occasionally visible as contours of thin batter traveling slowly around a spoon or fork during stirring. Complex flow patterns are encountered when the fluid contours are forced to change shape and/or dimension suddenly, as, for example, during passage through porous media (Lapasin and Pricl, 1995). At high velocities, turbulent flow results when molecules in any one plane travel with different translational and rotational velocities. Streamline and turbulent flow are characterized by a Reynolds number (2200), which is a dimensionless computation involving flow distances and time. Viscosity (η) is streamline below 2200.

An elongational or extensional viscosity (η_E) develops as a result of a conformational transition when disperse systems are forced through constrictions, or compressed or stretched (Kulicke and Haas, 1984; Rinaudo, 1988; Barnes *et al.*, 1989; Odell *et al.*, 1989; Clark, 1992). The intuitive logic is that the random coils resist the initial distortion. η_E is believed to elicit the human sensation of stringiness (Clark, 1995). If shear viscosity is denoted η_i, rheologists define a Trouton ratio as η_E/η_i wherein $\eta_E > \eta_i$ by a factor approximating 3 for uniaxial extension and 6 for biaxial extension. Alternatively stated, the Newtonian η_i calculates to one-third to one-sixth η_E (Steffe, 1992).

Barker and Grimson (1991) modeled the flow of deformable particles after a free-draining floc whose shape, orientation, and internal structure ranged between the extremes of an extended chain and a folded globule. They interpreted the unhindered motions of free-flowing, deformable droplets to result from an unbalanced force imposed by the flow field, resulting in rotations around the particles' center of mass; this rotation is superimposed on the steady translational motion.

If a fluid mass merely deforms under slight stress and the deformation (strain ϵ) is completely recoverable spontaneously upon removal of the stress, the fluid is an elastic fluid. Elasticity is sensitive to temperature and relatively insensitive to concentration, because of the already high solute content. Fluids exhibiting stages of viscosity and elasticity throughout the flow continuum are viscoelastic fluids. There are five transitional stages of viscoelasticity, viz., the glassy, leathery, rubbery, rubbery flow, and viscous stages (Cowie, 1991).

5. *Light Scattering and Turbidity*

The cloudy appearance of a dispersion seen when a beam of light impinges on it is the result of light scattering. Cloudiness is evidence that a boundary indeed exists between the liquid and solid phases. The conical rays observed as light emerges from a small aperture into dimly lit space—narrow at the point of incidence and wider away from it—are the well-known Tyndall effect. Light scattering is a function of colloidal particle size. Diameters less than 0.05 nm obey Rayleigh's law, stating an inverse mathematical relationship of wavelength (λ) to turbidity in isotropic solutions. Particles that have at least one dimension equal to or greater than λ undergo Debye scattering in which an angular dependence of the wave intensity is taken into consideration (Oster, 1960). Scattering is inversely proportional to the fourth power of the incident wave frequency: hence, dispersed polysaccharides scatter shorter λ much more than they do the longer λ, and violet waves are scattered 16 times as strongly as red waves (Smith and Cooper, 1957). Optically anisotropic polysaccharides scatter light more intensely at forward angles ($< 90°$) than at backward angles ($> 90°$). Scattering is also a function of concentration of the dispersed phase.

6. *Optical Activity and Anisotropy*

Sugar monomers contain asymmetric carbon atoms and are therefore optically active; i.e., they rotate plane polarized light. Optical activity diminishes with polymerization, and conversely increases with saccharification. The inherent asymmetry of polysaccharide molecules enables each one to be characterized by an axial ratio (ratio of cross-sectional and longitudinal distances normal to each other). Instrumentally, the relative intensities of light scattered at 45° and 135° have been rationalized into a dissymmetry coefficient (Z) capable of providing qualitative information about macromolecular dimensions (Stacey, 1956). Anisotropic molecules have axial ratio greater than 1. By phase microscopy, starch granules from different plant species can be identified through their distinctive polarization crosses.

7. *Surfactancy and Protective Colloid Action*

Incompatible commingling molecules separate soon after the commingling stimulus is withdrawn: such systems have short lifetimes and are therefore said to be unstable. For longer life, surface active compounds (surfactants), efficacious in small quantities, are added to decrease the contact angle between the immiscible surfaces, lower σ, and permit the interfusion of the immiscible surfaces.

The stabilizing function of macromolecular surfactants in solid–liquid systems is exercised through protective colloid action. To be effective, they must have a strong solution affinity for hydrophobic and hydrophilic entities. In liquid–liquid systems, surfactants are more accurately called emulsifiers. The same stabilizing function is exercised in gas–liquid disperse systems where the surfactants are called foam stabilizers.

Polysaccharides possess surfactancy to varying degrees. Methylcellulose is a very efficient protective colloid at 0.01% (Dow Chemical Co., 1990). Random coils (of xanthan) are more surface active than helices (Young and Torres, 1989).

D. Heterogeneity and Homogeneity

Isolated, purified polysaccharides are beset with numerous heterogeneities. To begin, structural and positional isomerism are caused by different locations and distributions of rhamnose, sulfate, and acetyl groups, for example, in the primary structure. Fine structures in galactomannans and pectins, consisting of smooth and hairy regions where proteins interact (Kravtchenko *et al.*, 1993), are sites of microheterogeneity. There is normally a wide spectrum of DP that arises from the use of chemical reagents in extraction processes. DP heterogeneity was denoted by Elias (1979) as polymolecularity and by Everett (1988) as polydispersity. The designation M refers to a monomolecular or monodisperse polymer, and $\overline{M}$ refers to a polymolecular or polydisperse polymer. Dimensional heterogeneity is typical of micelles and pores in a polysaccharide gel subjected to variations of the same physical treatment.

Size-homogeneous fractions of a polysaccharide can be separated from a polymolecular sample by a variety of separation techniques. Washing is an elementary, unintentional homogenizing operation, because lower DP fractions ($< 10^3$ Da) are apt to remain in the wash. Homopolymers are characterized by sharply defined boundaries and discontinuities that contrast with the diffuse boundaries of heteropolymers in the separation display. In natural and artificial heteroglycan synthesis, the distribution of a copolymer or fine structure may be in random, sequential, or block design, e.g., the galactomannans. Rhamnogalacturonan regions in pectin alternate with homogalacturonan regions (Kravtchenko *et al.*, 1993). Block copolymerization confers unusual bulk properties on linear polymers (Finkelmann and Jahns, 1989), and variations in copolymer composition alter the electrostatic environment and consequently the size, structure (Hunkeler *et al.*, 1992), and properties of polymers.

At moderate to high concentrations of polysaccharides, interactions over the length of the macromolecules beget an assortment of contact or junction zones (Rees, 1969) whose heterogeneous distribution is explained as follows (Silberberg, 1992): different segments are in different solubilities and hence different conformational states, with the result that some contacts are of longer duration than others. At high concentration, strong associations between the less soluble and more sterically complementary segments develop into crystallites with very long lifetimes. Junction-zone heterogeneity is widespread in gels.

Fluctuation theory explaining light scattering in dilute dispersions supposes transient heterogeneities initiated by density variations (Stacey, 1956).

E. Polymorphism, Hysteresis, and Syneresis

The polysaccharide primary structure affords numerous opportunities for hydroxyl- and carboxyl-group interactions, leading to polymorphism (allomorphism) and hysteresis. The difference between polymorphs is in the crystal unit-cell structure.

There are at least four crystalline forms of cellulose, based on different packing of the primary chain (Blackwell, 1982), and three forms of granular starch, based on the packing of double helices (Noel *et al.*, 1993). The differences are largely in the unit-cell dimensions and the crystallization and precipitation temperatures. One form of starch, precipitated with alcohol, is in a symmetrical molecular arrangement and is readily dispersible in cold water (Kerr, 1950). Mannan and dextran yield different crystals at low and high temperatures, and there was not only a polymorphic difference, but a conformational difference in cellulose (Quenin and Chanzy, 1987). Curdlan appears to have three polymorphs—anhydrous, hydrated, and annealed.

Hysteresis is the variable response of a system, e.g., a polysaccharide disperse system, to different processing and handling procedures, or the mode and order of application of a stimulus. The different procedures are referred to as pathways. Linear polysaccharides are prone to hysteresis traceable to different entanglements and disentanglements of the primary chain. A film made by hydrating a xerogel has a lower diffusivity than one made by dehydrating a hydrosol to an identical water content, because the former dispersion contains smaller pore diameters than the latter. The different hydration and dehydration pathways followed by the same dispersion earmark a phenomenon called sorption hysteresis. Similarly, rehydrated mashed potato has an eating texture different from that of freshly mashed potato at the same moisture content, because of hysteresis. Viscosity hysteresis arising from different treatment histories of the same disperse system can sometimes be a quality-assurance and quality-control problem. A hysteresis loop is shown for methylcellulose in Fig. 3, where two viscosities are observed at one temperature in a critical temperature range.

Mindful of the energy requirement for dispersion of a polysaccharide solute in water, syneresis is the slow, spontaneous separation of liquid from a gel, as the solid phase attempts to return to its energy ground state. This phenomenon is a quality defect, because it foreshadows solute sedimentation.

III. Phenomenology

Phenomenology is the study of behavior without explanatory structures and mechanisms. The chemically similar polysaccharides lend themselves to phenomenological study, because of the commonality of many of their properties and responses to ambient stimuli: for example, irrespective of their chemistry, polysaccharides are generally dispersible in water and indispersible in acetone; their η_i is concentration dependent within a critical range, and, with few exceptions, they respond identically to heat.

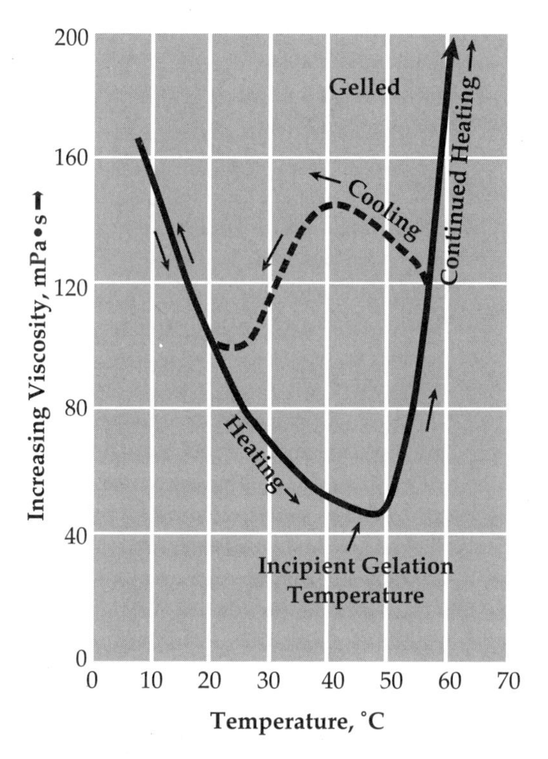

Figure 3 Thermal gelation of methylcellulose in aqueous solution (100 mPa s at 20°C and heating rate 0.25°C min^{-1}). Reprinted with the permission of the Dow Chemical Co., Midland, MI.

The phenomenology of amorphous cellulose and amorphous starch (amylopectin) transcends the conventional classification based on chemistry. Although opposites in branching and bonding, they hydrate equally well at room temperature and develop high η_i, whereas crystalline cellulose and crystalline starch (amylose) do not. Cellulose and starch crystallites possess other properties in common—more with each other than with their generic amorphous analogs. A carrageenan–water sol (polyanionic) is visually indistinguishable from a Konjac mannan–water sol (nonionic): each has the appearance of a thick paste at refrigeration temperatures.

IV. Property and Function Modifications

The DS, DE, and DP are intrinsic to the polysaccharide molecule, exercising influence on the solute's intra- and intermolecular interaction and interaction with water. Physical processes have extrinsic influences, insofar as they affect directly the DS, DE, and DP, and indirectly the dielectric property of water. Industrial suppliers offer a wide variety of DS, DE, and DP polysaccharides for diverse uses involving water solubility, dispersion stability, clarity, film strength, etc. A DS ≤ 0.1 is enough to modify the properties of starch (Dautzenberg *et al.*, 1994). The cold-water dispersibility, clarity, and stability of starch are best at DS = 0.02–0.2. The most widely used CMC has DS = 0.7 (Hercules, Inc., 1980). Some pectins do not gel with 65% soluble solids and acid outside a DS range of 43–85% (Pilgrim *et al.*, 1991). There are many ways to influence the polysaccharide response intrinsically through the DS, DE, and DP, and extrinsically through cosolutes and the solvent.

A. Acid

The 1,4-α- and 1,6-α-glycosides are stable for moderate intervals in acidic media at the strengths (pH 3.3–4.2) and temperatures (0–121°C) associated with processing low pH foods (e.g., fruit juices, jellies, sauces). With prolonged heating, reducing glycosides transmute and depolymerize in proportion to the severity of the process. The polysaccharide anionic character is strongly inverse to pH.

The β-glycosides, glycuronans, and pyranoses are more stable to acid than the α-glycosides and furanoses (Whistler and Smart, 1953; Marsh, 1966); 1,3-glycosides are unstable. Amylopectin is more acid-stable than amylose, identical in all but branching and anomeric configuration. Pectin is the relatively acid-stable hydrolysate of protopectin, its natural, acid-labile, insoluble precursor. Completely demethylated galacturonans (pectic acid) are yet more stable than methylated galacturonans and starch.

Starch is partially hydrolyzed under mildly acidic conditions to "thin-boiling" fluids that are less viscous and more translucent than the untreated starch dispersion, because of the reduction in DP. Strongly acidic media ($pH < 3.0$) promote more extensive lowering of the DP: acid-stable polyanions may precipitate directly without passing through the transitional gel state.

B. Alkali

Polysaccharides are differentially stable to cold dilute alkali, which facilitates classification as alkali-stable and alkali-sensitive (Reeves and Blouin, 1957). High concentrations of alkali swell fibrils, shift conformation, and depolymerize the primary structure. The 1,4-β bonds are more alkali-resistant than the 1,4-α bonds. Nonreducing glycosides are more stable than reducing glycosides. Ester and sulfate groups are easily hydrolyzed.

Polysaccharide β-elimination is a mild alkaline depolymerization reaction, whereby a molecule of water is introduced at the site of rupture where a double bond forms. An esterified hydroxyl (on C-6 of the monomer ring) is a prerequisite for β-elimination (Albersheim *et al.*, 1960). Metko and McFeeters (1993) studied the reaction and discovered that pectin with a higher degree of methylation degraded faster.

Amylose and amylopectin are easily solubilized by 1–2% NaOH. In an oxygen-free atmosphere, heated amylose isomerizes to saccharinates, but degrades rapidly in air; amylopectin degrades to its 1,6-α branch points. Advanced stages of alkalization (with heat) result in rearrangement, decomposition, and resynthesis to yellow–brown chromophores (caramel); the degradation is accelerated by phosphate and acetate (BeMiller, 1965a). Glycogen, the so-called animal starch, chemically similar to amylopectin, is not affected by 30% boiling NaOH, a concentration that destroys most biological tissues.

Cellulose is inert to all but the most drastic alkaline conditions. Mild alkali (0.5 N, 100°C) lowers the DP in time by progressively lysing the reducing-end monomer. It and concentrated alkali react heterogeneously[2] in numerous integrals of temperature, concentration, and time to give swollen fibers called alkali-cellulose that ultimately yields cellobiose, cellotriose, and cellotetraose (the soluble dimer, trimer, and tetramer, respectively, of glucose). In the terminal stages, numerous short-chain acids (Richards, 1963) appear in the reaction mixture. Cellulose homogeneous reactions recently have been made possible in *N,N*-dimethylacetamide–LiCl solvent (McCormick and Shen, 1982).

Partial demethylation enhances the physical reactivity of many natural polysaccharide esters. NaOH (0.1 M) is the reagent of choice for partial demethylation, with minimum impact on the primary structure. Deacetylated guar gum interacts more strongly with xanthan gum than does acetylated guar gum (Lopes *et al.*, 1992). Alkali causes galactomannans to become more coiled—a transition attended with a loss of η_i not attributable to depolymerization (Hui and Neukom, 1964). Alkaline pretreatment of κ-carrageenan enhanced the latter's gel-forming ability and raised the gel melting temperature (Watase and Nishinari, 1987).

At pH 14, cellulose ionizes to a polyanion (cellulosate) and the hydronium counterion (H_3O^+). Counterions are also called gegenions.

C. Oxidants

The controlled oxidation of otherwise stable polysaccharides increases their polarity. By oxidation in an alkaline medium, the C-6 hydroxyl group is converted to a carbonyl or carboxyl group, resulting in enhanced am-

2. In this context, heterogeneity refers to the occurrence of the chemical reaction in two phases—liquid (solvent) and solid (cellulose). The reaction is homogeneous if the two reacting phases are completely miscible.

phiphilicity and affinity for cations. The concentration of aldehydes, ketones, and acids depends on the severity and duration of the oxidation.

Oxidized starch loses its ability to gel, thereby making low η_i dispersions. Glucose dialdehyde is a glucose oxidation compound used to crosslink agarose (linear) and dextran (branched) for their performance as adsorbents. Oxidized celluloses are a substitute for glues manufactured from animal by-products.

D. Enzymes

The bond specificity of enzyme hydrolysis and the very mild reaction temperature not only ensure minimum degradation of polysaccharides, but safeguard against many of the secondary and side reactions associated with chemical hydrolysis. The choice of enzyme(s) has consequences for gel firmness and ion sensitivity. The initiation and progress of enzyme hydrolysis may require block or random sequences (Rexova-Benkova and Markovic, 1976; Ishii *et al.*, 1979). Enzymes from different sources may not be catalytic to related structures: for example, pectin, alginic acid, and gum tragacanth are three galacturonans not demethylated by the same esterase. The cation-binding capacities of enzyme-deesterified pectin differ from those of chemically deesterified pectin (Yalpani, 1988).

Enzyme activity may be enhanced by a chemical pretreatment, as in the example of alkaline hydrogen peroxide on corn fiber, where the hydrolytic rate was increased by a factor of 1.6 (Leathers, 1993). Some enzymes depolymerize polysaccharides by β-elimination.

Polysaccharides can be completely depolymerized by complementary enzyme action (Leathers, 1993): for example, starch is completely hydrolyzed sequentially to glucose, beginning with the amylases. α-Amylase excises amylose and linear side chains from amylopectin to yield a preponderance of maltotriose, maltotetrose, etc., and branched oligosaccharides called limit dextrins: the substrate η_i is significantly lowered. Without α-amylase, long segments are not converted to short-chain dextrins. β-Amylase (syn. diastase) reduces amylose and linear fragments to maltose. Neither enzyme can hydrolyze 1,6-α bonds, which require 1,6-α-glucosidase in whose absence the depolymerization terminates in branched oligosaccharides. In plant material, α-amylase is normally low and β-amylase is normally high. Commercial preparations are intentionally prepared to contain all the enzymes necessary for complete starch hydrolysis. β-Glycosidases hydrolyze the β-glycans.

In fermentations, the amylases are either present in or added to the polysaccharide substrate at activity levels sufficient to hydrolyze starch to maltose. Yeast does not generate the amylases, so malt (germinated grain) is relied on to augment their concentration. The importance of yeast is to produce maltase for converting maltose to glucose.

Glucoamylase depolymerizes starch, beginning at the nonreducing end, through the 1,4-α and 1,6-α linkages, terminating at (the theoretical) 100% glucose; thereby, liquid sweeteners with a high-dextrose equivalent[3] of 92–95 are obtained. Glucose isomerase isomerizes some of the glucose to fructose, yielding high-fructose (corn) syrups with yet greater sweetening power. Recent developments in carbohydrase technology include antistaling α-amylases, immobilized amylases, and amylases with increased tolerance of heat and acid (Hebeda and Teague, 1994).

E. Chemical Substituents

Mention has already been made of the numerous effects attendant upon chemical substitutions on the polysaccharide linear chain. Natural branches impart a dispersion stability to amylopectin that is not afforded amylose. One only has to compare cellulose ethers, deesterified chitin, and the lysis product of protopectin with the underivatized parent compound to appreciate the impact of chemical substituents on functionality. The loosening of compact, parallel structures with alkyl, hydroxyalkyl, and alkoxyl groups facilitates hydration and transforms insoluble, refractory polysaccharides to soluble, reactive polysaccharides. Not only do these substituents obstruct the crystallization tendency, they almost always confer secondary functionalities like η enhancement and foam, suspension, and freeze–thaw stabilization.

Whereas extracted, purified cellulose is comparatively physically inert, the carboxymethyl derivative is among the most functional of polysaccharides, and the hydroxyalkyl derivatives are among the most unusually alcohol tolerant. Propyleneglycol alginate does not gel because polypropylene glycol hinders the prerequisite ordering: this polyester is more acid stable than algin, a property attributed to bulky substituents (Wong, 1989).

Polysaccharide functionality is a variant of the nature and distribution of the substituents. A polar substituent makes a polysaccharide less hydrophobic, i.e., more hydrophilic, and vice versa. A uniform distribution of $-CH_2COO^-$ on the cellulose primary structure results in smooth, nongrainy nonthixotropic sols. The probability of nonuniform substitution increases at a low level of substitution of cellulose-containing crystallites, because the reagent is unable to penetrate the crystallite, regions (DuPont).

Deesterification of natural polysaccharides accomplishes the same result as chemical derivatization, to wit, the critical DE range necessary for high-methoxyl pectin gelation. Robinson *et al.* (1988) discovered that the normally gelling, branched polysaccharides studied did not gel when side chains were removed, because the unsubstituted main chain did not subsequently adopt an ordered conformation. Their explanation was that the polydispersed

3. Dextrose equivalent refers to sugar concentration reported as dextrose, i.e., 100 (mg reducing sugar per milligram of dry substance).

linkages might still have prevented ordering. The acetyl group found naturally in many polysaccharides is gel-inhibiting at concentrations above approximately 2.6% (Pippen *et al.*, 1950). Chitin, an acetylated aminocellulose, is far less reactive than chitosan, the deacetylated derivative.

F. γ-Radiation

Polysaccharides are degraded by ionizing radiation, with crosslinking as a possible side reaction, via free radical mechanisms depending on the water content (Bellamy and Miller, 1963). Radiolysis occurs at all doses, sometimes with an initial increase in η, as more polysaccharide becomes solubilized (King and Gray, 1993). Low doses are an effective means of obstructing aggregation.

γ-Radiation has been shown to be a feasible way to convert higher DP alginates to lower DP alginates with improved properties (King, 1994). Loss of η accompanying such depolymerizations can transform pseudoplastic fluids to Newtonian fluids. In carrageenans, radiolysis was shown to be initially rapid and then to decrease to a constant, low, radiation-insensitive DP: the rate was faster in solution than in the solid state (Marrs, 1988).

G. Micromolecules

The solvent quality of water deteriorates by additions of alcohol and salt, singly and combined. Polyanions are especially sensitive to cations and are invariably precipitated by the di- and polyvalent species in sufficiently high concentration. Charge suppression by nonsolvents and cations decreases the polyanionic coil volume (v_c) and simultaneously reverses the solute–solvent interaction through θ to precipitation. Micromolecules can alter the dispersion rheology (Pastor *et al.*, 1994) of charged an neutral polysaccharides.

1. Salt

Electrolytes affect dispersed polysaccharides through water inactivation, specific ion binding, and polyanion neutralization. Each effect is valence-dependent, but is less on neutral polysaccharides than on ionic polysaccharides. Di- and polyvalent cations gel or precipitate a constant amount of polysacchride at much lower concentrations than do monovalent cations. The precipitation reaction is used to advantage in isolating pectin with alkaline Al^{3+}, because this cation and polymeric forms of $Al(OH)_3$ readily precipitate and entrain pectinic acid from apple tissue homogenates. Other di- and polyvalent cation effects are crosslinking (Prud'homme *et al.*, 1989) and an increased rate of β elimination over monovalent cations (Sajjaanantakul *et al.*, 1993). Dilute concentrations of any of the valences can have a "salting in" or stabilizing effect on mostly dispersed polyanions; conversely, high concentrations can have a "salting out" or destabilizing effect. The polysaccharide sensitivity varies with the DS, DP, concentration, temperature, pH, and the order of additions of components to the solvent (Hercules, Inc., 1980). NaCl at 0.6% doubled the viscosity of 1% CMC in water when added after instead of before dispersion of the CMC (Stelzer and Klug, 1980). NaCl has virtually no effect on xanthan at any concentration in the practical use range (0.10–1.0%); in starch dispersions, this concentration level depresses gelatinization and raises the gelatinization temperature (Howling, 1980). The effect of NaCl on starch gelatinization is complex (Lund, 1984). NaCl lowers the phase-separation temperature of aqueous hydroxypropylcellulose (Hercules, Inc., 1971). Sodium κ- and ι-carrageenan do not gel, but the potassium salts do. Potassium CMC and sodium CMC are water-soluble, but aluminum, ferrous, and ferric CMC are insoluble. Salt augmented the pH influence on the helix–random-coil transition in curdlan (Ogawa *et al.*, 1972, 1973b); in gellan, it reduced helix aggregation at low temperatures and reduced coil dimensions at high temperatures (Miyoshi *et al.*, 1994). Salt accelerated the rate of crystal formation in starch (Mita, 1992).

2. *Sugar*

The simplest explanation of the effect of sugar in sufficient concentration on polysaccharide dispersions is that it lowers the water activity and initiates hydrophobic interactions (Morris, 1985). In jelly-making with pectin, Michel *et al.* (1984) assign to 65% sugar the role of creating a poor solvent in which pectin molecules stiffen prior to aggregation. Stated differently, sugar competes with dispersed polysaccharides for water (Howling, 1980). Sugar also lowers the phase-separation temperature of hydroxypropylcellulose dispersions (Hercules, Inc., 1971), heightens certain intrinsic viscosity functions, stabilizes the ordered structure of gellan (Crescenzi and Dentini, 1988) and carrageenan gels (Nishinari and Watase, 1992), increases the rupture strength of carrageenan gels (Fiszman *et al.*, 1986), raises the starch gelatinization temperature, and narrows the gelatinization range (Eliasson, 1992). The order of raising the gelatinization temperature is sucrose > glucose > fructose (Bean *et al.*, 1978).

Sugar is claimed to do more than simply lower water activity (Lund, 1984): it increases the number and size of junction zones in agarose gels (Nishinari *et al.*, 1992). κ-Carrageenan setting and melting temperatures were shifted upward with increasing concentrations of sugar (Nishinari and Watase, 1992).

Edible plant tissues, notably of fruits and vegetables, may be saturated with sugar by a process called vacuum infusion, whereby tissues are immersed in the sugar solution and subjected to a vacuum emission of air from the intercellular space: the sugar solution replaces air. The occluded sugar retards the spontaneous deterioration that cellulose would otherwise undergo as a result of the dehydration and packing of microfibrils. Calcium and/or pectinmethylesterase may be included in the sugar solution to enable simultaneous, cooperative binding of pectinic acid after demethylation of the carboxyl groups (Poovaiah and Moulton, 1982; Javeri *et al.*, 1991).

Retail packs of pectin for jelly-making utilize sugar as a diluent, so that the pectin does not cake and is suspended in water as discrete particles at the outset of mixing.

H. Homogenization

Homogenization[4] is the uniform reduction in particle size by shearing, as a dispersion or suspension is forced through micron-sized apertures: simultaneously, viscosity and buoyancy are enhanced, thus prolonging the life of the otherwise unstable dispersion. Shearing can rupture chemical bonds and result in a high surface-to-charge ratio: this operation can convert a hydrated cellulose suspension to a hydrogel (Walter *et al.*, 1977).

V. Volume and the Theta Condition

The volume relationship under θ (V_θ) and non-θ (V_i) conditions is stated as

$$V_i = \alpha V_\theta. \tag{1.2}$$

α is called the expansion factor. In a good solvent, $\alpha > 1$; it decreases in numerical value as the solvent becomes less good. The ideal condition for measuring intrinsic molecular properties is at θ where $\alpha = 1$ and coil dimensions are unperturbed, i.e., independent of the influence of solute-solute and solute-solvent interactions. The unperturbed state is difficult to reach by dispersed polysaccharides, given their spontaneous disposition to aggregate.

VI. Summary

Polysaccharides are mostly beneficial polymers, but a number of them may be deleterious in some situations. Their properties and utilitarian value reside less in their chemical constitution than in their ability to change

4. A mechanical process unrelated to homogeneous solvent processes.

shape, size, properties, and function in numerous responses to ambient stimuli: water, cosolute, and heat exert and strongest influences. They are prone to self-association. The nature and magnitude of the responses, irrespective of chemistry, make them amenable to phenomenological study.

The 1,4-α-polysaccharide random coil is a disordered, high-energy distribution of independently rotating segments, except when rotation is hindered by bulky substituents, side chains, and unfavorable surroundings. Given all the possibilities of rotation and interaction, it is logical to conclude that polysaccharide conformation is a dynamic property, constitutively based on the DP in a minor way, but largely dependent on and highly susceptible to ambient stimuli. Substituents and side chains modify many useful properties of refractory polysaccharides: deesterification of natural derivatives is sometimes beneficial.

CHAPTER **2**

The Polysaccharide–Water Interface

I. Introduction

Water, occasionally modified by ethanol and salt, is the universal food solvent. V_i and v_m are maximum in pure water and minimum in pure ethanol. Without hydration, polysaccharides cannot perform as plasticizers, thickeners, texturizers, stabilizers, crystallization inhibitors, and bulking and gelling agents. Crosslinking holds v_m constant, irrespective of the solvent. All water relationships with solute theoretically disappear at θ.

Three polysaccharide–water interfaces are shown in Fig. 1. The agarose interface is observed to be indifferent to water (Hayashi and Kanzaki, 1987), judging from its distinctly sharp, hydrophobic boundary. The methylcellulose and pectin boundaries are diffuse and clearly hydrophilic. The air bubbles on methylcellulose attest to its efficacy as a foam stabilizer. The interfaces depicted in Fig. 1 can be visualized on a macromolecular scale as the interfaces between dispersed polysaccharide molecules and water.

There are at least four forces in effect at the polysaccharide–water interface, viz., van der Waals, ionic, hydrogen bonding, and hydrophobic (Ilmain *et al.*, 1991): van Oss (1991) adds a repulsive force due to Brownian motion. van der Waals forces are always attractive, because they depend on oscillating nuclei and electrons of paired molecules (Adamson, 1990); ionic and hydrophobic forces are attractive or repulsive, depending on the paired species; hydrogen bonding is attractive and paramount, because of the mutual affinity of hydroxyl and carboxyl groups. The strength of the hydrogen bond is 10–40 kJ mol^{-1}, intermediate between van der Waals (1 kJ mol^{-1}) and ionic (500 kJ mol^{-1}) bonds (Israelachvili, 1992). Urea is a hydrogen-bond breaker.

The fundamental concept of a polymer molecule in solution is that it consists of a number of segments, each approximating the size of a solvent

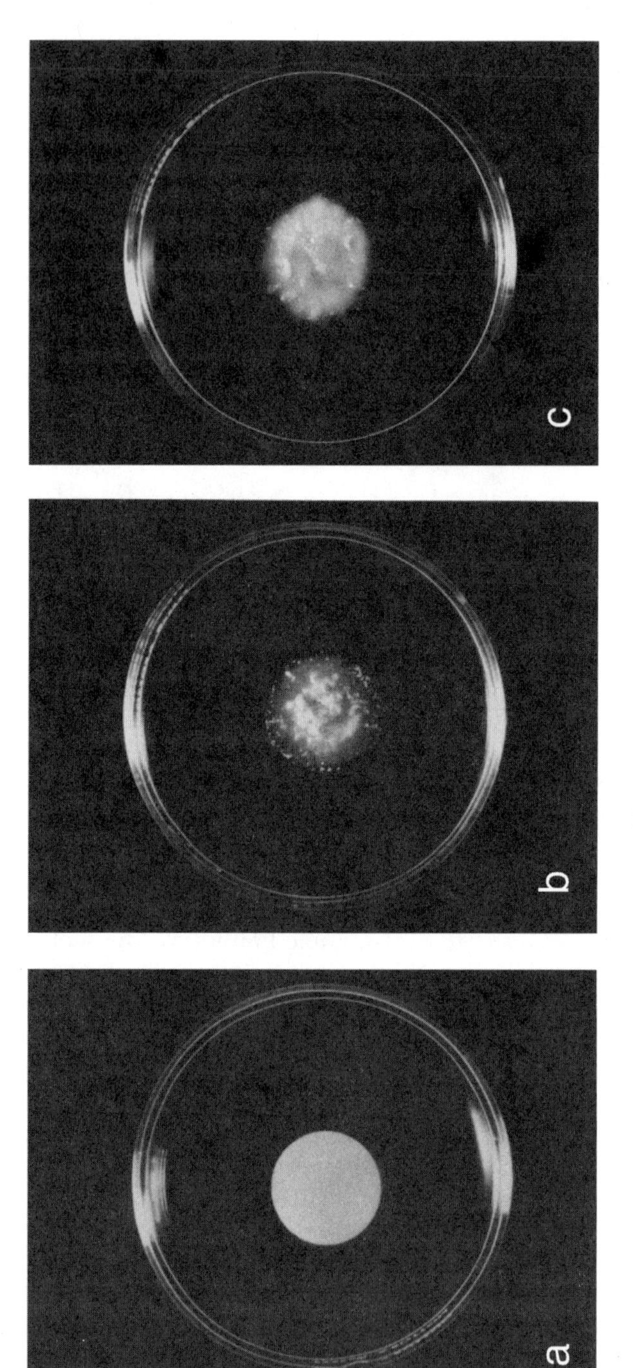

Figure 1 Interface between water and an agar (a), methylcellulose (b), and pectin (c) gel. The agar (a) was prepared by pouring 1.5% hot sol into a 1-in.-diameter plastic die and allowing it to cool. Methylcellulose and pectin (b and c) were prepared similarly. The dies were then immersed in ethanol and placed in Petri dishes in preparation for photographing. The photographs were taken after 6 months (a), 24 h (b), and 4 h (c). Note the sharp boundary in (a), the adhering air bubbles in (b), and the diffuse interface in (c).

molecule (Flory, 1953). The attractive forces at the segment–solvent interface are stronger by far in an aqueous polysaccharide dispersion than in a solution of a synthetic polymer and an organic solvent. The forces generated by contiguous segments are described as short-range forces; these are more likely repulsive than attractive and therefore tend to expand v_m. Long-range forces are generated by interaction between distant segments, and are repulsive and attractive, but more likely to be attractive and therefore to lessen v_m. Short- and long-range forces are collectively called the excluded volume effect, because they outline a discrete volume for each molecule—the excluded volume (v_{ex})—from which all other molecules are barred. The v_{ex} effect is claimed to be the only important interaction between uncharged polymers (Dautzenberg *et al.*, 1994). By convention, the v_{ex} effect is positive when the net force is repulsive and is negative when the net force is attractive.

II. Properties of Water

The properties of H_2O obtain from the tetrahedral geometry of the molecule. The hydrogen atoms form an angle of 104.5° with the strongly electronegative oxygen atom, and the tightly bound electrons induce an asymmetric charge distribution with a center of negative charge at the oxygen site and a center of positive charge at each hydrogen: this electronic configuration confers 40% ionic character on the molecule (Belitz and Grosch, 1986). In the pure state, H_2O has four other H_2O molecules attached to it (coordination number of 4).

A. The Dielectric Constant

H_2O calculates to a dielectric constant (D_o) of 78.54 at 25°C, the highest of the ordinary solvents. Ethanol has $D = 24.3$ at 25°C. Given its high polarity, H_2O easily engages in dipole–dipole and ion–dipole reactions.

B. Ionization

At neutral pH and 25°C, one water molecule in every 10 million becomes ionized through the bonding of one of its protons to another water molecule (Baianu, 1992a); West and Todd (1961) estimate one molecule in every 550 million. In practical terms, every 1 g mol (18 g) of pure H_2O renders 10^{-7} g mol of hydrogen ion (10^{-7} g of H^+) and 10^{-7} g mol of hydroxyl ion (1.7×10^{-6} g of OH^-). In aqueous solution, H^+ is hydrated to H_3O^+ (the hydronium ion). H_2O is 10 times more ionizable at 60°C than at 25°C,[5] and 40 times more ionizable at 100°C (West and Todd, 1961). There is evidence that H^+ exists also as $(H_9O_4)^+$ (Baianu, 1992a). In any of its cationic forms, H_2O engages in ion–ion and ion–dipole reactions.

C. Activity

The concept of a chemical potential is germane to a discussion of water activity (a_w), which is technologically defined as the ratio of the equilibrium water vapor pressure over a solution or dispersion (p_o) and the water vapor pressure over pure water (p_o^0). Also by definition, the chemical potential of a solvent (μ_o) or a solute (μ_i) is the rate of change in energy of either with a change only in the molal content of that component in solution.

According to Raoult's law (Williams *et al.*, 1978),

$$\mu_o = \mu_o^0 + RT \ln(p_o/p_o^0). \tag{2.1}$$

μ_o is the chemical potential of the solvent in the vapor phase, μ_o^0 is the standard chemical potential (potential of pure solvent), and p_o/p_o^0 is a vapor–pressure function proportional to the mole fraction of solvent. In a dilute polysaccharide dispersion behaving ideally it follows that

$$\mu_o = \mu_o^0 + RT \ln[n_o/(n_o + n_i)]. \tag{2.2}$$

n_o and n_i are the number of moles of solvent (o) and solute (i), respectively, and $n_o/(n_o + n_i)$ is the mole fraction ($\bar{n}_o$). Correspondingly, $\bar{n}_i$ is the mole fraction of solute (i) and in a binary dispersion, $\bar{n}_o + \bar{n}_i = 1$ always. In terms of the mole fraction of solvent, Eq. (2.2) may therefore be rewritten as

$$\mu_o = \mu_o^0 + RT \ln(1 - \bar{n}_i). \tag{2.3}$$

At pure-water equilibrium, $\bar{n}_i = 0$ and

$$\mu_o = \mu_o^0, \tag{2.4}$$

$$\mu_o/\mu_o^0 = a_w = 1. \tag{2.5}$$

Whereas for a solution of small molecules, $\ln(1 - \bar{n}_i)$ is an increasingly large negative number as $\bar{n}_i$ increases, μ_o is progressively lower than μ_o^0 as micromolecular solute is added. For aqueous polysaccharide sols, $\bar{n}_i$ is exceedingly small and $1 - \bar{n}_i$ remains effectively constant in a large volume of water.

5. Calculated from data tabulated in *Handbook of Chemistry and Physics*, 62nd ed., CRC Press, Boca Raton, FL, 1981–1982.

TABLE I
Water Activity (a_w) of a Selection of Hawaiian Starchy Foods at Two Different Temperatures[a]

Foodstuff	a_w 26.7°C	a_w 6.7°C	Moisture (%)
Poi[b]	1.00	0.86	74.6
Kulolo[c]	1.0	0.84	55.7
Rice cake[d]	0.98	0.84	48.0
Manapua[e]	0.97	0.88	35.9

[a] Walter and Seeger, 1990.
[b] Steamed and mashed taro.
[c] Baked taro and coconut pudding.
[d] Steamed rice flour confection.
[e] Steamed pork-filled bun.

The stability of a moist food is in proportion to the magnitude of a lowering of a_w from 1. With the exception of lipid auto-oxidation, it is only at $a_w = 0.2–0.4$ that the chemical (e.g., flavor changes, reaction rates) and biological (e.g., microbial growth, enzyme catalysis) stability of moist foodstuffs is assured. Lipid autooxidation rates are maximized by a reduction in a_w, because of the enzyme's heightened activity in proximity to the substrate. Hydrosols do not show this low level of a_w; dried fruits show $a_w = 0.72–0.80$ (Belitz and Grosch, 1986). From these facts, it is inferred that polysaccharides do not impart biological stability to liquid foods by a lessening of a_w; refrigeration is necessary (Table I), T being the alternative variable in Eq. 2.3.

If n_o and n_i are expressed as their partial molal volumes ($\bar{v}_o$ and $\bar{v}_i$), the total volume of dispersion (V_i) is

$$n_i/(n_o + n_i) = (n_i\bar{v}_i)/[(n_o\bar{v}_o) + (n_i\bar{v}_i)], \tag{2.6}$$

$$(n_o\bar{v}_o) + (n_i\bar{v}_i) = V_i, \tag{2.7}$$

$$n_i/(n_o + n_i) = (n_i\bar{v}_i)/V_i. \tag{2.8}$$

n_i/V_i is the solute molarity (c_m) and

$$n_i/(n_o + n_i) = c_m\bar{v}_i. \tag{2.9}$$

At equilibrium, the solvent and solute chemical potentials are equal, i.e.,

$$\mu_o = \mu_i, \tag{2.10}$$

Eq. (2.2), in terms of component i, may then be written

$$\mu_i - \mu_i^0 = RT \ln c_m \bar{v}_i \,. \qquad (2.11)$$

D. Specific Heat

The quantity of heat necessary to raise the temperature of a substance 1°C is the heat capacity of that substance. The specific heat capacity (synonymous with specific heat) of the substance is the heat capacity divided by its mass (weight). The specific heat capacity of water is 1 cal g^{-1} $°C^{-1}$, among the highest of solvents. Because of this, aqueous systems heat and cool relatively slowly. An infinitesimally slow rate of heat exchange is a necessary condition for thermodynamic reversibility. If water is cooled infinitesimally slowly without disturbance, it supercools; i.e., it persists in the liquid state well below its normal freezing point. The sudden imposition or withdrawal of a stimulus at the temperature of supercooling causes the supercooled water to revert to ice. In contrast, polysaccharide sols are supercooled by rapidly lowering the temperature with stirring.

E. Contraction and Expansion

At 4°C, 1 g of water has minimum volume (1.003 cm^3) and maximum density (0.997 g cm^{-3}). The coefficient of volume expansion above 4°C to the boiling point (100°C) is 2.1×10^{-4} cm^3 $°C^{-1}$—among the smallest in solvents. At a specific volume (reciprocal density) of 1 cm^3 g^{-1} and molar mass of 18 g, the volume fraction (ϕ_o) is quite large relative to that of a dispersed phase.

F. Surface Tension

Due to σ, water droplets completely surrounded by air assume a spherical shape. Sphericity not only offers the lowest ratio of surface area to volume, it also indicates a strong, uniformly distributed, unbalanced tension, directed toward the center, which must be overcome in any surface expansion. The spherical geometry of water droplets thus complies with the law governing entropy, stating a spontaneous minimization of surface energy, and hence surface area, at equilibrium. If σ is the energy necessary to expand or contract the surface area by 1 cm^2, its energy unit is elaborated into ergs per centimeter squared or dynes per centimeter.

The σ of ethanol is 22.3 dyn cm^{-1} and that of pure water is 71.97 dyn cm^{-1} at 25°C: the latter is the highest or ordinary solvents, varying only slightly with temperature. Any ethanol additions to water (o) therefore lower σ_o. The σ of solids is measurably less than that of liquids.

At 25°C, sucrose raises σ_o from 72.50 dyn cm^{-1} at 10% concentration to 75.70 dyn cm^{-1} at 55% concentration; at 20°C, sodium chloride does so, from 72.92 dyn cm^{-1} at 0.58% concentration to 82.55 dyn cm^{-1} at 25.92% concentration.[6]

Nonequilibrium σ gradients can be the driving force behind water migration from phase to phase or surface to surface. Marangoni flow (Ross and Morrison, 1988) is such transport from a region of low σ (weak cohesive forces) to one of high σ (strong cohesive force). Nonequilibrium σ gradients can result from temperature, concentration, thermal and compositional differences, and compressions and dilatations of adsorbed films at interfaces.

III. Polysaccharide–Water Interactions

Hydrocolloidal water is an integral part of the dispersed phase and travels at the same velocity with it; this is considered tightly "bound" water. Yakubu *et al.* (1990) identified three other forms of water in potato and corn starch, viz., weakly bound, surface trapped, and bulk water. All forms were not present in potato starch containing less than 35% moisture, but were present in corn starch. Water, far removed from the solute surface (unbound or free water in the outer volume), travels at a different rate from hydrocolloidal water (Lechert *et al.*, 1981).

Most polysaccharides affect the mobility and structuring of water (Blanshard, 1970) beyond the immediate interface (Barfod, 1988) to a thickness of several molecular diameters (Rickayzen, 1989). The reciprocal effect on flowing polysaccharides of surface-water immobilization and structuring is a contribution to distortions from sphericity, and hence to flow birefringence.

Where there is no solute–solute association, macromolecules may act simply as a viscosity enhancer of the continuous phase: Barnes *et al.* (1989) call this phenomenon neutral interaction. Through what is called hydrodynamic interaction (Dautzenberg *et al.*, 1994), the streamlines of hydrocolloidal particles flowing past each other affect each other. Tightly bound water apparently does not contribute much to a_w (Yakubu *et al.*, 1990). Free water is removable from a sol by freezing, while simultaneously, soluble trace components concentrate in the hydrocolloidally bound, unfrozen water, often to saturation.

In a binary dispersion, there is usually one of five interfaces to consider, where a polysaccharide, for example, may act as a protective colloid. These interfaces are liquid–solid (sol), liquid–liquid (emulsion), solid–solid (mixed xerogel), liquid–air (foam), and solid–air (powder). In any of these systems at

6. *Handbook of Chemistry and Physics*, 62nd ed., CRC Press, Boca Raton, FL, 1981–1982.

equilibrium, $\sigma_{o,i}$ is the difference between the higher and the lower component σ and, consequently, at a water–polysaccharide interface, $\sigma_{o,i}$ is always positive (Jirgensons and Straumanis, 1962):

$$\sigma_{o,i} = \sigma_o - \sigma_i . \tag{2.12}$$

The function of a protective colloid is to lower $\sigma_{o,i}$ to a minimum. In practical language, wetting is an attempt by a surfactant to accomplish this by lowering the contact angle, which enables liquids to spread over each other, on its mission to make the phases mutually miscible. Relative to solvent and component, the concentration of a protective colloid is quite low, but it accumulates at the interface, theoretically as a thin film. Micromolecules that wet surfaces dissolve completely in the solvent. One unique property of micellar surfactant electrolytes is their ability to solubilize some otherwise insoluble organic molecules (Adamson, 1990).

With exceptions, amphiphiles lower $\sigma_{o,i}$ and inorganic electrolytes raise it (Jirgensons and Straumanis, 1962: Hiemenz, 1986). Long-chain fatty acids and alcohols lower $\sigma_{o,i}$ by decades as a function of their concentration. Traube's rule (Jirgensons and Straumanis, 1962) states that fatty acids lower σ_o in proportion to the chain length. Large concentrations of sucrose raise $\sigma_{o,i}$ only slightly (Vold and Vold, 1983). Of the polysaccharides with varying degrees of surfactancy, methylcellulose is claimed to be the most efficacious; xanthan gum is more so as a random coil than as a helix (Young and Torres, 1989). The glycomannans are particularly adept at the prerequisite organization for surfactancy into lamellar liquid crystals at an oil–water interface (Reichman and Garti, 1991).

IV. Influences on Polysaccharide–Water Interactions

Four features of a polymer solute figure prominently in the polysaccharide–water interactions, viz., bonding, branching, ionization, and nonuniformity of the repeating structure (Glass, 1986).

A. Bonding

With mixed results, correlations have been attempted between polysaccharide composition and function. Beginning with simple sugars, sucrose (an α-D-glucopyranosyl–β-D-fructofuranoside) and maltose (an α-D-glucopyranosyl–α-D-glucopyranose) are truly soluble; cellobiose (a 1,4-β-D-glucopyranosyl–β-D-glucopyranose) and the trimer raffinose (1,4-α-D-galactopyranosyl–1,6-*O*-α-D-glucopyranosyl–1,2-β-D-fructofuranoside) are not. The 1,4-α- and 1,6-α-D-glycosides contain the most hydrophilic linkages, but the 1,4-β-glycoside correlates with insolubility and crystallinity. Inulin, a biopolymer of 1,2-β-D-fructofuranose, is soluble. The 1,3 bonding provides less symmetry than 1,4 and 1,6 bonding, and consequently less chain–chain associations, which ultimately enables better dispersibility. 3,6-Anhydro-α-L-galactose is structural to the gelling of carrageenans.

B. Branching

Amylopectin and glycogen, differing only in the frequency and length of branching at the sixth carbon of the glucose monomer, are readily hydratable, but amylose, the linear counterpart, is not. Guar gum, a gelling polysaccharide with many uniformly spaced α-D-galactopyranosyl monomers in the smooth region of the fine structure, has a higher water affinity than does nongelling locust bean gum with fewer unevenly spaced α-D-galactopyranosyl monomers in the hairy regions. Scleroglucan, a 1,3-β-glucan substituted with 1,6-β-glucose, does not normally gel; yet curdlan, with a similar primary structure but without most of the glucose substituents, does gel. In the view of Yalpini (1988), the fact that the least branched galactomannans develop gels upon standing suggests that artificial branching does not improve hydration.

C. Ionizing Groups

Ionization offers dispersed polyanions short-term protection from deposition through shielding with H_3O^+ counterions. The hydration of xanthan is enhanced by its containing a charged, trisaccharide side-chain repeating unit (Sanofi, 1988).

D. Heterogeneity

Glass (1986) conceived of irregularities in a polysaccharide chain as possibly promoting dissolution. He cited the carrageenans as one example, admitting, however, that experience to date is inadequate to predict polysaccharide behavior based on chain heterogeneity. It is noteworthy that guar and locust bean gums (both heteropolysaccharides) are compatible with the widest spectrum of other polysaccharides.

V. Polysaccharides as Adsorbents

Micromolecules and ions, initially dissolved in the outer, free-draining volume of a solution interfacing a solid surface including a polysaccharide surface, accumulate by diffusion (osmotic migration) across an imaginary, semipermeable membrane into the inner, adsorbed layer of water on the surface. For most solid–liquid systems, the accumulation theoretically ceases at a monolayer or equilibrium concentration (plateau value). At equilibrium, μ_i is equal on both sides of the membrane. This conceptual behavior, called positive adsorption, is reminiscent of a concentration cell in which the diffusion rate of the migrating solute is proportional to the concentration (c_i). Negative adsorption (desorption) is the reverse process. The Gibbs adsorption [Eq. (2.13)] for dilute solutions formalizes positive and negative adsorption of nonelectrolytes:

$$\partial c_i = -(RT)^{-1} c_i \cdot \partial \sigma_o / \partial c_i \,. \tag{2.13}$$

∂c_i is the excess concentration on the solid adsorbent surface per unit cross section, and $\partial \sigma_o / \partial c_i$ is the rate of change of σ_o with c_i. The greater the surfactancy in the bulk solution, the lower is $\partial \sigma_o$ and the higher is ∂c_i. As a consequence of positive adsorption, the layer of water on the solid surface passes from high (initial state) to low (final state) σ, $\partial \sigma_o / \partial c_i$ becomes negative, and ∂c_i becomes positive. As Eq. (2.13) shows, the accumulation is inversely temperature-dependent, and the rate of change of surface tension with concentration is directly temperature-dependent.

The surface of a powdered polysaccharide equilibrated in air is hydrophobic and resistant to wetting—a condition that poses difficulty in dispersal when cereal flour, for example, is mixed with water in the preparation of doughs and batters. Dispersion usually requires a large expenditure of mechanical energy.

VI. Polysaccharides as Adsorbates

Equilibrium or monolayer adsorption of a polysaccharide as adsorbate is unlikely, except in the latter process, as a result of chemisorption, whereby valence forces extend to no more than one molecular distance. Instead, the first layer of polysaccharide provides an adsorption site for the second layer, *ad infinitum*, in a nonequilibrium process, until phase inversion. Macromolecules including polysaccharides do not desorb: they accumulate in multilayers with an increased rate of adsorption at higher temperatures.

The Freundlich equation, empirical in origin, relates positive adsorption to a power function of c_i, as follows:

$$w/g = k' c_i^{1/k}, \tag{2.14}$$

$$\log(w/g) = 1/k(\log c_i) + \log k'. \tag{2.15}$$

w/g is the weight of adsorbate per gram of adsorbent. The range of k is 0.1–0.5 (Daniels *et al.*, 1970). The validity of Eq. (2.15) is proven when $\log(w/g)$ vs $\log c_i$ is a straight line at a constant temperature. When $k = 1$, the rate of surface accumulation is equal to the rate of change in solution concentration, and at equilibrium, there is a 1:1 distribution of c_i between the surface and the solution. When $k > 1$, the adsorbent is highly efficient at accumulating c_i; the opposite holds for $k < 1$. Equation (2.15) is most adequate to quantify adsorption of small electrolytes from solution over a considerable range of concentrations (Glasstone and Lewis, 1960). An important characteristic of Eq. (2.15) is the limiting value of w/g as k increases.

The Brunauer–Emmett–Teller equation governing multilayer adsorption shows inflections above monolayer saturation as the adsorbate accumulates on a surface over an extended interval. Heller (1966) factored in the time variable:

$$t/(w/g) = k + k't. \tag{2.16}$$

w/g is now the weight of adsorbate per gram of adsorbent at any time t, k' is the slope, and k is the intercept. The inflections in polymer adsorption isotherms are explained by Adamson (1990) as possibly deriving from physical adsorption over a chemisorbed layer, which results in the observed isotherm being the sum of two isotherms. Another explanation is that different surfaces that have different adsorbabilities exist on the adsorbent.

As adsorbate, the polysaccharide ∂c_i is sensitive to the DP and polydispersity (Cohen Stuart *et al.*, 1982). Higher DP polysaccharides are less kinetically active, are therefore slower to accumulate than lower DP polysaccharides because of the time taken for surface orientation, and are thus more inclined to stay adsorbed longer and reach higher concentrations. Agitation increases the rate of physical adsorption. From the foregoing discussion on polysaccharide dispersibility, it is safe to conclude that multilayer adsorption is antecedent to polysaccharide phase inversion and in some instances to sol–gel transition.

Chemisorption is irreversible adsorption, which suggests valence bonding at specific sites on a surface. Transition metal ions, protein below its isoelectric point (positively charged), and di- and polyvalent cations are prone to chemisorption.

As an adsorbate, a polysaccharide is modeled as orienting itself linearly in flattened conformations (Dickinson and Euston, 1991a), any of which may engage it in stabilization and destabilization. If only a segment is adsorbed, the remainder protrudes into the bulk liquid as tails, loops, and trains (Van de Ven, 1989): these localized conformations may be restricted to the interfacial layer or may migrate from it (Lips *et al.*, 1991).

Galactomannans, particularly, become adsorbed and organized on an oil-drop surface as lamellar liquid crystals that perform as steric and mechanical barriers to coalescence (Reichman and Garti, 1991). The surfactancy of xanthan was found to be related to the amount adsorbed (Young and Torres, 1989).

VII. Summary

Polysaccharides interfaced with water act as adsorbents on which surface accumulations of solute lower the interfacial tension. The polysaccharide–water interface is a dynamic site of competing forces. Water retains heat longer than most other solvents. The rate of accumulation of micromolecules and microions on the solid surface is directly proportional to their solution concentration and inversely proportional to temperature. As adsorbates, micromolecules and microions ordinarily adsorb to an equilibrium concentration in a monolayer (positive adsorption) process; they desorb into the outer volume in a negative adsorption process. The adsorption–desorption response to temperature of macromolecules—including polysaccharides—is opposite that of micromolecules and microions. As adsorbate, polysaccharides undergo a nonequilibrium, multilayer accumulation of like macromolecules.

CHAPTER 3

State- and Path-Dependent Properties

I. Introduction

A system approaching thermodynamic equilibrium in infinitesimally slow steps is characterized by reversibility of the steps without a change in the energy status of the infinitesimal transitions. At equilibrium, the integral of the variables is constant, so that a change in one limits change in the others. The integral is a state function, and the associated properties are state properties. If an integral of identical variables depends on the pathway, the properties are path-dependent and inexact. Sorption hysteresis, supercooling, and diffusivity, for example, are inexact properties.

The thermodynamics of small molecules is predicated on equilibrium states arrived at in relatively short time intervals, making them essentially time-independent. Polysaccharide processes seldom reach thermodynamic equilibrium in a practicable time interval.

II. Mass–Volume–Pressure–Temperature Relationships

One mole of gas (n) has is volume (V) specified by temperature (T) and pressure (π) according to the Charles–Boyle law, and any change in either of these two variables results in a corresponding change in the other, so that the following equation is satisfied:

$$n = \pi V/RT. \tag{3.1}$$

This equation is a mathematical statement that T/V is constant in any constant-pressure process and T/π is constant in any constant-volume process. Macromolecules in very dilute aqueous concentration imitate gas molecules, in conformance with Eq. (3.1). In binary polysaccharide dispersions, conformational transitions of the dispersed phase introduce negligible volume change (ΔV) in the dispersion; however, the solvent phase (water) experiences noticeable ΔV at vaporization at 100°C and freezing expansion at 0°C.

III. Electrostatics and Electrokinetics

A pair of polysaccharide molecules approaching each other in water exerts an interaction potential (ξ') that is the algebraic sum of the competing attractive and repulsive forces. ξ', integrated over all pairs of molecules, is ξ. This principle is embodied in the Derjaguin–Verwey–Landau–Overbeek (DLVO) theory of colloidal stability (Ross and Morrison, 1988). The equilibrium distance between the molecules is related to c_i, the volume of the hydrated particles, ionic strength, cosolute, nonsolvent additions, temperature, and shearing.

A. Nonionic Polysaccharides

Neutral molecules, dissolved, dispersed or suspended in a liquid medium, are in continuous random motion (Brownian motion) with a mean free path (x) and collision diameter (x_e), depending on c_i and v_{ex} effects. At a far separation distance, ξ is negative, increasing to 0 at x_e, where repulsion counterbalances attraction and the amphiphiles are at dynamic equilibrium in a primary minimum energy state. At $x < x_e$, the molecules repel each other and ξ is positive. High concentrations shorten x and make the collision rate nonlinear with c_i (Hammett, 1952). A separation distance of $x < x_e$ is sterically forbidden without fusion.

Coulomb's law of electrostatic attraction between two unlike charges (Q_1 and Q_2) states

$$F = Q_1 Q_2 / \mathrm{D_o} x^2. \tag{3.2}$$

F is the force of attraction (Coulombic attraction), seen to be inversely proportional to the second power of x. Given the closer proximity of Q_1 and Q_2, small-diameter particles experience a much greater force of attraction than large-diameter particles. Between a point charge and a dipole, F varies with x^{-3}, and between an ion and a dipole, F varies with x^{-2} (Adamson, 1990). Coulombic attraction between ions of opposite charge is of much longer range than van der Waals attraction between neutral molecules (Alberty and Silbey, 1992). The latter acts through 6–7-nm distances, decaying exponentially with the sixth or seventh power (Van de Ven, 1989) of x.

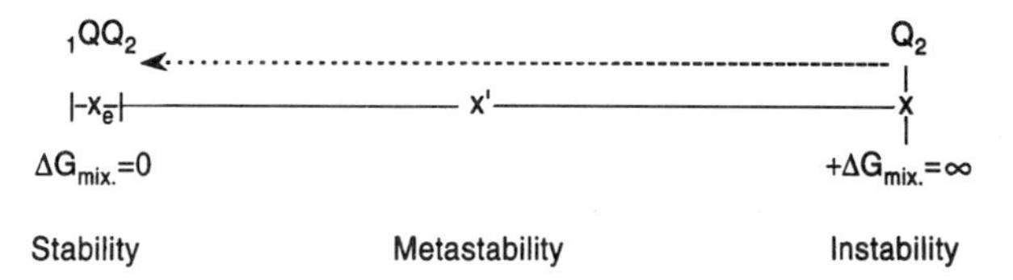

Figure 1 Diagram of the correspondence between the distance (x) and stability of a polysaccharide dispersion containing solute particles (Q) in water. As the particles approach each other (Q_2 approaching Q_1), the interparticle distance x' is shortened and the dispersion passes through stages of metastability to x_e, the collision diameter. For a freshly prepared sol, the particles experience the highest Brownian activity and the increase in energy is maximum ($+\Delta G_{\text{mix}} = \infty$), whence it declines to the primary energy state at equilibrium, when $\Delta G_{\text{mix}} = 0$. Aging is the process of $+\Delta G_{\text{mix}} = \infty$ declining to $\Delta G_{\text{mix}} = 0$.

This inverse power integer is called the Born exponent (Gould, 1962). Simple power relationships (Israelachvili, 1992) apply to neither polysaccharide intermolecular nor surface forces due to the multiple effects of substituents, branches, kinks, etc.

The total energy content (E) of an aqueous polysaccharide dispersion is stated in the general equation

$$E = \int_0^x F\,dx = Q_1Q_2/\mathrm{D_o}\int_0^x dx/x^2 = -Q_1Q_2/\mathrm{D_o}(1/x + k). \tag{3.3}$$

The negative sign indicates a loss of energy up to x_e as Q_1 approaches Q_2. Absolute energy states are difficult to measure, but more importantly, differential (d) and integral (Δ) changes are quantifiable, once a reproducible reference state has been decided on. Numerical changes are computed by subtracting initial assigned values from final assessed values. Although the law of entropy stipulates that molecules spontaneously strive to descend from a metastable E (at any x') to a primary minimum (at x_e), it is the goal of the food researcher to cause the kinetically active molecules to exist almost indefinitely in a metastable, kinetically active state at any $x' > x_e$ (Fig. 1).

B. Ionic Polysaccharides

It used to be thought that cations simply precipitated polyanions, but it was recognized later that electrolytes had special valence and solvent-mediated effects on a hydrosol other than neutralization of opposite charges (Holmes, 1922). It is now firmly established that ionization of the carboxyl and sulfuric acid groups in ionic polysaccharides, or adsorption of ions on neutral macromolecules, is an initial step in electrokinetic mechanisms of stabilization and destabilization.

1. Dissociation

In the normal pH range of living tissues, weak carboxyl and sulfuric acid groups connected to biopolymers partially dissociate in water and maintain equilibrium with an equivalent weight of H_3O^+ whose concentration is highest at the surface of the polyanions and decreases outward. The ionizing group may be integral to the primary structure as in pectin or pendant to it as in CMC. The degree of dissociation is directly related to the pH of the dispersion medium. In excess H_3O^+ or other counterions, ionization is depressed and the macroanion then behaves like its neutral counterpart (Pals and Hermans, 1952) or as does a salt of the polyanionic acid. The acid strength is indicated by the dissociation constant (K_z) (Table I): the larger the K_z, the stronger the acid. Interestingly, hyaluronic acid (not listed; $pK = 3.23$; Cleland *et al.*, 1982), containing equimolar quantities of β-D-glucuronic acid and *N*-acetyl glucosamine, is among the strongest of the biological organic acids. Hyaluronic is an important compound in animal physiology, but is not intentionally added to food.

TABLE I
The pH and Ionization Constant (K_z) of Some Washed, Aqueous Dispersed Ionic Polysaccharides ($\leq$ 1%)[a]

Polysaccharide	pH	K_z
CMC (0.50%)	7.0	$10^{-4.5}$
CMC (0.75%)	7.0	$10^{-4.5}$
CMC (1.0%)	7.0	$10^{-4.5}$
Gellan	7.2	$10^{-4.4}$
Gum arabic	5.1	$10^{-5.9}$
LM pectin[b]	3.2	$10^{-4.2}$
LM pectin[b]	3.1	$10^{-4.0}$
Algin	6.2	$10^{-4.2}$
HM pectin[c]	3.2	$10^{-3.9}$
HM pectin[c]	3.6	$10^{-4.3}$
AM pectin[d]	4.4	$10^{-4.9}$
AM pectin[d]	4.3	$10^{-5.1}$
Xanthan	7.0	$10^{-4.2}$
Carrageenan	7.0	$10^{-4.9}$
Agar	6.7	$10^{-4.3}$

[a] Walter and Jacon, 1994.
[b] Different grades of low-methoxyl pectin (Hercules, Inc., Wilmington, DE).
[c] Different grades of high-methoxyl pectin (Hercules, Inc.).
[d] Different grades of amidated pectin (Hercules, Inc.).

The identicalness of the ionization sites in a linear polyelectrolyte (Tanford, 1961) stimulated the interest of Walter and Jacon (1994) in a possible relationship between K_z and $\overline{M}$ of ionic polysaccharides displaying the characteristic titration curve of a weak, monobasic acid. Without any theoretical assumption, Eq. (3.4) was derived from simple algebra by combining elementary principles of the dissociation theory of weak acids with polymer segment theory:

$$\overline{M} = [10^{-\text{pH}^*}]/\{[10^{-2\text{pH}} + 10^{-(\text{pH}^* + \text{pH})}]f' \cdot f''\}. \quad (3.4)$$

A segment factor f' related $\overline{M}$ to the constant ratio between the carboxyl group and the monomer. With specific reference to pectin, a second factor (f'') was later postulated to account for the degree of carboxylation.[7] Subsequently, a series of pectins determined to have an anhydrogalacturonan content of 68–96% ($f'' = 0.68$–0.96) did not always change the exponent of $\overline{M}$. A CMC sample with a 0.75 degree of carboxylation, making $f'' = 0.25$, increased the exponent by 1.

Offering only theoretical surmises, Cesãro and Villegas (1996) refuted the $\overline{M}$–pH relationship, without first addressing important aspects of it. Charge density and distribution, and location (pendant on ring) of the ionizing group, are probable factors to be considered.

2. *The Electric Double Layer*

The solution arrangement of ionic polysaccharides as a primary layer of polyanions (the Stern layer) and a secondary layer of H_3O^+ counterions creates a model of an electric double layer (also called the Helmholz double layer). The number of and average distance between adjacent charges on the polysaccharide surface are defined as the charge density. An imaginary shear plane separates the inner volume containing the fixed primary layer and the outer volume containing diffusely distributed H_3O^+ (and other cations present). This shear plane is estimated to be within a few molecular diameters of the particle surface (Hiemenz, 1986). Along this shear plane, a frictional coefficient (f_c) is generated that is a property of the size, number, and distribution of segments. A potential (the zeta potential, ζ)[8] develops between the inner and outer volumes where, in the latter, $\zeta = 0$. The higher the charge density in the primary layer, the higher is ζ, the more stable is the double layer, and, consequently, the more stable is the dispersion. Ions migrating with water across the shear plane in the direction of the polyanion lower ζ and initiate destabilizing events.

7. Three carboxylation sites on a glucose monomer.

8. The reader is referred to Vold and Vold (1983) for the difference between the zeta and Stern potentials.

The swelling of plant parenchyma tissue has been ascribed to an electric double layer, which depends significantly on the type of counterions that constitute it: monovalent ions cause more swelling than do divalent ions (Shomer *et al.*, 1991), probably because of their larger ionic radius.

The electric double layer has been compared to a capacitor (Jirgensons and Straumanis, 1962)—a single device for storing positive and negative charges separately. An electrical potential is created by the charge separation across a narrow space where an inserted dielectric becomes polarized, effectively augmenting Q/ξ (defined as capacitance C), although the quantity of $+Q$ and $-Q$ is not changed. The augmentation results from cancellation by the dielectric's positive end of an equivalent fraction of $-Q$ and by the negative end, of an equivalent fraction of $+Q$, thus effectively lowering ζ and raising C. The dielectric constant of a substance is defined as the ratio of Q/ξ, measured with the substance inserted in the space between the charged leaves of the capacitor and Q/ξ measured *in vacuo*. In an ionic polysaccharide dispersion, water is the dielectric and C is the double-layer stabilizer.

Electrolytes affect ξ and the charge distribution. The fundamental equations (Jirgensons and Straumanis, 1962) are

$$\zeta = kQx/\mathrm{D_o}, \tag{3.5}$$

$$x = k'(\mathrm{D_o}/i_s)^{0.5}. \tag{3.6}$$

i_s is the ionic strength and x here is the double-layer thickness or Debye length, an important property of the double layer (Cabane *et al.*, 1989). For some hydrocolloids destabilization occurs below a critical ζ. In the DLVO theory, the Debye length is the distance over which ξ falls to $1/e$ or 37% of its value at the surface. The "salting in" and "salting out" of polyelectrolytes are related to i_s and x. At a critical i_s, $x = 0$. Equations (3.5) and (3.6) are in harmony with the common observation that water (high $\mathrm{D_o}$) normally stabilizes and ethanol (low D) destabilizes an ionic polysaccharide dispersion. Alternatively stated, $(C)_{\mathrm{water}} > (C)_{\mathrm{ethanol}}$.

Equation (3.6) confirms what is already known from experience: Ca^{2+} and Al^{3+} have greater flocculating power than Na^{1+} and K^{1+}. In the other direction, polysaccharide polyanions are less inclined to disperse in water containing multivalent ions than in water containing monovalent ions.

3. *Electrophoresis, Electroosmosis, and Streaming Potential*

Given the charge, polysaccharide polyanions are electrically conducting in sols and gels, and move in an electrical field (electrophoresis) in compliance with the equation

$$\Omega = k(\xi/x)\zeta\mathrm{D_o}/\eta_o. \tag{3.7}$$

η_o is the solvent viscosity, Ω is the electrophoretic mobility, small for macroions relative to microions, and ξ is the applied electromotive force across x. Neutral polysaccharides do not migrate in an electrical field, except as a moiety of an ionic complex or when they adsorb ions. Electrophoresis is a useful method for studying heterogeneity (Aspinall and Cottrell, 1970).

In electroosmosis, solvent travels with the charged species and the volume (v) migrating per unit time in a standard cell is

$$v = D_o \zeta \xi / \Lambda \eta_o . \tag{3.8}$$

Λ is the conductivity of the solution. A streaming potential ζ is established by a confined solution flowing under pressure through small-diameter pores and capillaries. It is believed that the confining walls, typically glass, become charged with OH^-, thereby initiating the potential.

4. The Donnan Distribution

Diffusion across a semipermeable membrane causes an increment of Na^+ (y'), for example, to migrate from the outer solution containing Na^+Cl^- at concentration y toward the inner solution containing the polysaccharide polyelectrolyte salt (Na^+P^-) initially at c_i. The migrating Na^+ is accompanied by an equal concentration of Cl^- to maintain electrical neutrality. At equilibrium, Na^+ and Cl^- on the P^- side are $c_i + y'$ and y', respectively, and the quantity of each in the outer volume is $y - y'$:

$$y'(c_i + y') = (y - y')^2, \tag{3.9}$$

$$y' = y^2/(c_i + 2y). \tag{3.10}$$

As shown in Eqs. (3.9) and (3.10), the concentration of migrating cations is a squared function of the adsorbate concentration in the outer volume of the solution. The Donnan distribution is a source of serious error when determining $\overline{M}$ of ionic polysaccharides by membrane osmometry. Polyanions may be freed of cations and excess H_3O^+ by electrodialysis.

IV. Thermodynamics

Classical thermodynamics discourses equilibria in solutions of noninteracting micromolecules, within and between phases, under constant and variable concentration, volume, temperature, and pressure. Deviations observed with macromolecules necessitated new theory modeled after a lattice in which segments of linear macromolecules, instead of the whole molecule, interchange positions with identical segments and with solvent molecules without bond rupture. Each segment, first estimated at about 25 chain atoms long (Powell and Eyring, 1942), is configured by its immediate neighbors to which it is bonded. The minimum length that a freely rotating polysaccharide segment can be is the monomer.

A. Enthalpy

The first law of thermodynamics (enthalpy) expresses the equivalence and interchangeability of the different forms of energy (heat, work, etc.), so that a molar mass of polysaccharide, for example, undergoing transformation from A to B, absorbs or evolves an increment of energy (ΔE) expressed (Glasstone and Lewis, 1960; Knight, 1970) as

$$\Delta H = \Delta E + \Delta(pV) - T\Delta S. \tag{3.11}$$

ΔH[9] is the total energy exchange, ΔE is the internal energy change, and ΔS is the change in entropy (*vide infra*). $\Delta(pV)$, having dimensions of work (force times distance),[10] is the energy expended in the transition, e.g., nonrecoverable pV work during viscous flow. At constant pressure, $p\,\Delta V$ is solely the work of expansion. Wunderlich (1990) adds an extra term to Eq. (3.11) to account for $p\,\partial V$ necessary to promote a conformational change of random-coil polymers. A negative ΔH indicates an exothermic process and a positive ΔH indicates an endothermic process. Under isothermal conditions, no heat is gained by the dispersion or lost to the surroundings, all exchangeable energy is ΔH, and ∂V is accompanied by a corresponding change in ΔH [$\partial(\Delta H)$], i.e.,

$$\partial(\Delta H) = p\,\partial V. \tag{3.12}$$

For polysaccharide dispersions, ∂V is exceedingly small relative to V_i. Equations (3.11) and (3.12) are mathematical propositions that the exchangeable energy stored in a dispersed polysaccharide solute is equal to the energy absorbed from an external source and any increase in surface area of the solute is consequently a repository of $+\Delta E$. Conversely, aggregation and desorption correspond to a loss of energy, felt as heat in the latter occurrence ($-\Delta E$) when a dry polyaccharide powder is wetted (positive adsorption).

In a nonisobaric coil–stretch transition,

$$\Delta H = n_i RT \ln(p_A/p_B), \tag{3.13}$$

9. $\Delta H = (H_o - H_i)_A - (H_o - H_i)_B$.

10. pV equals force per centimeter squared times centimeters cubed, which cancels to force times centimeters.

where p_A and p_B are the respective equilibrium pressures at states A and B. At constant V_i and p, ΔH is at once equated with its mechanical equivalent, the heat capacity ($C_{p,v}$):

$$\Delta H = C_{p,v}\,\Delta T. \tag{3.14}$$

ΔT is measured between an initial temperature (T_A) and a final temperature (T_B). Experimentation has shown that ΔH of polymer solutions is not very large, depending only weakly on T (Allcock and Lample, 1981). One mole of water has $\Delta H = 0.99828$ cal at 25°C and 1 atm.

The Clausius–Clapyron equation [Eq. (3.15)] is of incidental interest, because it states the molar relationship of water in a dispersion in equilibrium with its vapor—the definition of a_w [Eqs. (2.1)–(2.5)]:

$$\log(p_B/p_A) = \pm(\Delta H_{\text{vap}}/2.303R)(\Delta T)^{-1}. \tag{3.15}$$

ΔH_{vap} is the heat of vaporization of water.

B. Entropy

A descriptive definition of entropy (S) is that it is the amount of energy in a system unavailable for exchange (Glasstone and Lewis, 1960; Knight, 1970). A change in entropy (ΔS) from state to state is the practical thermodynamic variable for indexing the extent of randomness. ΔS is a function of the number of polymer segments per unit volume of dispersion and of the distance from a surface or from each other (Van de Ven, 1989). The most kinetically active macromolecules are the most randomly dispersed and possess the highest ΔS. Random coils are high ΔS conformations (Poland and Scheraga, 1970) vis-à-vis rods and helices. $-\Delta S$ is typical of sol-to-gel and amorphous-to-crystalline transitions. In an isothermal, reversible physical event such as solution of a micromolecule in water,

$$\Delta H = T\,\Delta S. \tag{3.16}$$

ΔS is related to $C_{p,v}$ as follows:

$$\Delta S = n_i C_{p,v}\ln(T_B/T_A). \tag{3.17}$$

C. Free Energy of Mixing

For the spontaneous merger of two phases to occur, the following condition must prevail:

$$G_{\text{mix}} \leq (G_{\text{solvent}} + G_{\text{solute}}), \tag{3.18}$$

where $G_{mix} \leq G_{solvent} + G_{solute}$ are, respectively, the Gibbs free energy of a solution, solvent, and solute. Spontaneous dissolution requires that ΔG_{mix} be negative. The dissolution of polymers including polysaccharides must almost always overcome a condition of $G_{mix} > (G_{solvent} + G_{solute})$.

In terms of ΔH and ΔS,

$$\Delta G_{mix} = \Delta E + \Delta(pV) - T\Delta S_{mix}, \tag{3.19}$$

$$\Delta G_{mix} = \Delta H_{mix} - T\Delta S_{mix}, \tag{3.20}$$

where ΔH_{mix} equals $H_{solvent}$ minus H_{solute} and ΔS_{mix} equals $S_{solvent}$ minus S_{solute}. The unit of ΔH, ΔS, and ΔG_{mix} is Joules per Kelvin.[11] The θ temperature is that temperature where $\Delta H_{mix} = 0$ and polymer dissolution exhibiting ideal behavior is instigated by ΔS only.

By itself, ΔS_{mix} is incapable of predicting spontaneity and randomness; this is demonstrated in crystallization and helix formation that anomalously result in a high degree of order $(-\Delta S)$, but are nevertheless spontaneous processes more significantly driven by a loss of latent heat $(-\Delta H)$.

For some constant-temperature, physical processes, e.g., phase separation and sedimentation, there is a corresponding $\partial(\Delta G_{mix})$ for every ∂p:

$$\partial(\Delta G_{mix}) = V\,\partial p. \tag{3.21}$$

Contemporary polymer theory considers segments of the primary structure to be the statistical unit comparable in size to that of solvent molecules. The large number of segments in polymers and the small scale of ΔG_{mix}, ΔH_{mix}, and ΔS_{mix} allow their thermodynamics to be preferably described statistically (Smith, 1982), thereby permitting the following equations:

$$\Delta G_{mix} = RT(X_o n_o \phi_i + n_o \ln \phi_o + n_i \ln \phi_i), \tag{3.22}$$

$$\Delta H_{mix} = RTX_o n_o \phi_i, \tag{3.23}$$

$$\Delta S_{mix} = -R(n_o \ln \phi_o + n_i \ln \phi_i). \tag{3.24}$$

X_o is a positive, inverse temperature-dependent interaction parameter per solvent molecule (Allcock and Lampe, 1981).

The Boltzmann law computes to a configurational ΔS governed by Eq. (3.22). A configurational ΔS represents dissolution of a perfectly ordered, pure solid polymer in pure solvent (Allcock and Lampe, 1981). van Oss (1991) cautions against designating physical processes as ΔH- or ΔS-driven unless careful microcalorimetric measurements have been made, because many thermodynamic suppositions (imputed to modeling or intuition) have not been substantiated by experimentation. Although descriptive analyses of

11. One joule equals 0.239 cal.

ΔH, ΔS, and ΔG may help to elucidate a mental picture of events, the thermodynamic status of most polysaccharide dispersions depends on the processes it underwent to achieve a desired end-product or use.

D. Irreversible Thermodynamics

Nonequilibrium, time-dependent processes manifest mostly as transport and relaxation (Wunderlich, 1990) describe polysaccharide events. The terminal outcome of transport is a change in position (potential energy), and that of relaxation is restoration exactly or approximately to an initial energy state. Because polysaccharide events are nonequilibrium processes, the addition or subtraction of energy is necessary for reversibility.

V. Kinetics

The inordinately long intervals required for many polysaccharide events to approach equilibrium necessitate that time and rate changes be more useful considerations. Kinetics is the study of the rate at which molecules arrive at or depart from an equilibrium state. The kinetics of large molecules is governed by the modification of rate laws applied to micromolecules.

A. Diffusion

The unidirectional diffusion of solute is a function of cross-sectional area (A) and time along a concentration gradient ($\partial c_i/\partial x$):

$$\varsigma = -AD(\partial c_i/\partial x), \tag{3.25}$$

$$D = (RT/\mathbf{N})(1/6\text{pi}\,\eta_o r). \tag{3.26}$$

ς is the amount diffusing, D is the diffusion coefficient (diffusivity) in dimensions of distance^{-2} time^{-1}, $\mathbf{N}$ is Avogadro's number, r is the particle radius, and pi = 3.142. The negative sign indicates decreasing concentration at the point of origin. Equation (3.25) is known as Fick's first law of diffusion. The flux of the solute is the rate of change of ς with time across A. In reality, D is itself a function of c_i and possibly $\partial c_i/\partial x$ (Geddes, 1949).

Equation (3.26) is adapted to nonspherical particles by multiplying D by a dissymmetry factor (Geddes, 1949).

D is related to the frictional coefficient f_c of a macromolecule, given by (Williams *et al.*, 1978)

$$D = RT/\mathbf{N}f_c \tag{3.27}$$

and to ΔH by (Hannay, 1967)

$$D = D_A \exp(-\Delta H/RT). \tag{3.28}$$

D is inversely related to the incidence of junction zones in hydrogels and directly related to the uniformity of the distribution of pores (Silberberg, 1989). Major difficulties in applying Eqs. (3.25)–(3.28) to polysaccharides arise from their polymolecularity and v_{ex} effects, because molecules of different sizes diffuse at different rates and interactions preclude the fundamental independent motion implicit in Fick's first law.

B. Order of Reactions

A binary dispersion of polysaccharide and water is effectively unimolecular, given the constancy of c_o and ϕ_o. Such a system is represented by

$$c_1 = c_0 e^{kt}, \tag{3.29}$$

$$2.303 \log(c_1/c_0) = kt. \tag{3.30}$$

k is the rate constant. When $c_1 = 2c_0$, Eq. (3.30) reduces to

$$t_{0.5} = 0.693/k. \tag{3.31}$$

$t_{0.5}$ is the length of time (half-life) required for 50% of any quantity of solute to accumulate (+) or be dispersed (−), in a first-order process. At true equilibrium,

$$\Delta G_{\text{mix}} = n_i RT \ln(c_1/c_0). \tag{3.32}$$

n_i is the number of moles of i distributed between c_1 and c_0.

Starch gelatinization—a unimolecular occurrence—follows pseudo-first-order kinetics after an initial time lag (Okechukwu and Rao, 1996a).

VI. Hydrodynamics

Similar to Eq. (1.2), the v_c of a polysaccharide random coil is related to v_θ through α:

$$v_c = \alpha v_\theta \, . \tag{3.33}$$

Above θ, $\alpha > 1$. As θ is approached, solvent departs the interior of the coil (Hiemenz, 1986), the initially non-free-draining water becomes progressively indistinguishable from the free-draining water, all interactions cease, and $\alpha = 1$. θ is hardly ever achieved in polysaccharide dispersions, considering the numerous opportunities for intra- and intermolecular bonding of –OH and –COOH groups that drive the spontaneous tendency to lessen the surface area by aggregation. A free-draining macromolecule is conceived as being in a more extended than coiled conformation, thereby permitting water to flow in streamline along the surface, unencumbered by adsorbed water. The superimposition of charge on the coil complicates the polyanionic response, which is then a function of the charge density and the uniformity of the distribution.

A. The Imaginary Shear Plane

The solute–water interaction extends 1–3 nm (Israelachvili, 1992) and decays exponentially with distance (Van de Ven, 1989). Non-free-draining water is water within this distance traveling with the same velocity as the particle nucleus. At the interface between the non-free-draining (bound) water and the outer volume of free-draining water traveling at a different velocity, an f_c [Eq. (3.27)] is generated. In this sense, hydration and the imaginary shear plane have enormous ramifications for human oral sensations elicited by dispersed polysaccharides.

B. The Equivalent Hydrodynamic Sphere

The classical thermodynamic and kinetic model is that of a rigid sphere impenetrable by water. A spherical geometry has been observed in many polysaccharide systems, notably hyaluronic acid–protein complexes (Ogston and Stainer, 1951), dispersed gum arabic (Whistler, 1993), and spray-dried ungelatinized starch granules (Zhao and Whistler, 1994). Spherulites of short-chain amylose were obtained by precipitation with 30% water–ethanol (Ring *et al.*, 1987), and spherulites of synthetic polymers were obtained during the initial stages of crystallization (Khoury and Passaglia, 1976). To accommodate topologically linear macromolecules, the concept of an equivalent hydrodynamic sphere, responding identically to external stimuli as a linear molecule, was introduced (Tanford, 1961). The polysaccharide model is that of a random coil approximating the peripheral outline of a sphere—the equivalent hydrodynamic sphere. The radius of this sphere is the hydrodynamic radius (R_h).

VII. Free Volume

Early theory propounded the existence of holes in a liquid that accommodated flow, as molecules "jumped" from hole to hole (Eyring, 1936). Modern theory perceives spaces in a polymer melt originating from randomly distributed segments of the primary structure, whose cooperative bond rotation (crankshaft motion) creates "free volume" (v_f), thus enabling the polymer chain eventually to achieve new positions. For a gram of dispersed solute, v_f is the difference between the specific volume of solute (v_{sp}) and v_{ex}:

$$v_f = v_{sp} - v_{ex}. \qquad (3.34)$$

VIII. Temperature Dependence

T is the defining parameter of both the thermodynamic and kinetic states of polysaccharide dispersions. With declining T, η increases and v_f decreases until $v_f = 0$ when all macromolecular motion ceases and the dispersion becomes essentially "frozen" with the onset of brittleness. With increases in T, the hydrogen-bond strength decreases, a poor solvent may become good, and a good solvent may become better. The strength of the hydrophobic bond increases with T (Ben-Naim, 1980).

According to the law of distribution of molecular velocities (Glasstone and Lewis, 1960), molecules in two different phases, at equilibrium, are related in translation through the Boltzman equation, stated as

$$n_1/n_2 = \exp(\pm E/RT). \qquad (3.35)$$

n_1 and n_2 are the molar concentrations in phases 1 and 2, respectively. The exponential factor $\exp(\pm E/RT)$ is the Boltzman factor indicating an exponential change in motion from phase 1 to phase 2.

In most polysaccharide sols, phase changes like gelation, gelatinization, and melting show an inflection on graphs of the logarithm of η versus T^{-1}. The gelation temperature (T_{gel}) may or may not be a function of the rate of cooling, which is apparently variable at high rates (Hinton, 1950) and constant at low rates (Walter and Sherman, 1983). Some substituted celluloses display anomalous behavior in that η is directly proportional to T, to T_{gel}.

Heating results in an apparently permanent loss of η without bond rupture in guar and CMC (Rao *et al.*, 1981). Henderson (1988), who observed the lowering of T_{gel} in the methylcellulose dispersion with increases in c_i, concluded that the thermal gelation was by dewatering due to hydrophobic bonding of the methyl groups.

The rubbery, sol–gel transition of an amorphous polymer eventuates over a narrow interval rather than at a single temperature. The midpoint of this interval is the glass transition temperature T_g (Kaelble, 1971; Cowie, 1991; Levine and Slade, 1992), where $v_f = 0$. A perfectly crystalline polymer does not possess a T_g; instead, it becomes completely ordered or disordered at a precise critical temperature—the melting temperature T_m. Melting is described as a first-order transition, because of the exactness of the heat of fusion. T_m is a state property. Naturally occurring, extracted microcrystallites are imperfect as a result of their matrix associations with a miscellany of other molecules, and therefore do not show a T_m, but more accurately show a softening temperature range variable with c_i, the DP, prior history, rate of cooling, crystal-size homogeneity, and water of hydration. T_m is located at higher temperatures for higher c_i and is constant at high DP. A single-bonded linear polymer experiences complete freedom of vibrational and rotational motion above T_m where the flow response is purely viscous. The greater flexibility of a random coil causes its T_m to be lower than for other conformations. From T_m to T_g, long segments may "freeze," whereas short segments remain mobile—a characteristic of viscoelasticity. For all these natural polymers, T_g is path-dependent and $T_m > T_g$.

From absolute zero (0°K) to 25°C, most hydrophilic solute remains separated in water to an upper critical solution or upper consolute temperature (T_c) (Glasstone and Lewis, 1963) whereupon they merge. In the opposite direction (from high to low temperature), solute and solvent or two solute phases in a common solvent may remain separated to a lower T_c, where they again merge. Many cellulose derivatives have a lower T_c in the vicinity of 45°C. The lower and upper T_c are called cloud points because of the incipient cloudiness observed there. This incipient cloudiness in a formerly translucent dispersion is evidence that the solute has emerged from a secondary minimum on its way to a gel (Walstra *et al.*, 1991).

Cloudiness may be induced at a constant T by nonsolvent and electrolyte additions. Electrolyte criticality is 10–100 times more effective from mono- to trivalence. The counterion dependency is known as the Schulz–Hardy rule. At a constant cation monovalence, the flocculation value (a critical minimum concentration of electrolyte usually in millimoles per liter) varies directly with the hydration radius. Arranged in decreasing order of effectiveness, the flocculating power of food electrolytes is $Mg^{2+} > Ca^{2+} > Na^{+} > K^{+}$ (the Hofmeister or lyotropic series). The correlation with anions is less certain, but for a common cation, $OH^{-} > Cl^{-} > -OSO_3^{-}$ (Jirgensons and Straumanis, 1962).

IX. Rheology

At 1-atm pressure in the surroundings, polysaccharide deformation and flow are normally initiated either by gravity or an applied shear rate ($\dot{\gamma}$); solvent (water) only flows under temperature (T) and concentration (c_i) gradients. When η_i is constant or independent of the rate of shear ($\dot{\gamma}$ in s^{-1}) or stress (τ), the flow is Newtonian. Very dilute polysaccharide dispersions are characterized mostly by Newtonian flow. At moderate concentrations, η_i may decrease (shear-thinning; synonymous with pseudoplastic) or increase (shear-thickening; synonymous with dilatant) nonlinearly with $\dot{\gamma}$: for these dispersions, η_i is replaced with η_a (the apparent viscosity). Low DP and uniform distribution of substituents are conducive to η_i; high DP and nonuniform distribution are conducive to η_a. A high η_a is believed to elicit the human oral sensation of "thickness."

In shear-thinning fluids at constant T, η_a decreases with increases in $\dot{\gamma}$ and the tertiary structure is instantaneously and completely recoverable upon return to $\dot{\gamma} = 0$. At very low values of $\dot{\gamma}$, shear-thinning fluids exhibit constant η_a (zero-shear viscosity). Shear-thinning can also occur under constant τ with the passage of time. $\dot{\gamma}$ has the opposite effect on shear-thickening fluids. A shear-thinning fluid is thixotropic if the declining η_a slowly returns to its initial value after $\dot{\gamma}$ has ceased. Similarly, the initial structure returns to a shear-thickening fluid after relief from $\dot{\gamma}$. Structural heterogeneity contributes to thixotropy, because the dispersed particles are able to interlock into semipermanent structures.

The slower and faster rates of η_a changes in shear-thinning and shear-thickening fluids result from molecular disentanglements and entanglements, respectively, wrought sometimes by slow stirring (low $\dot{\gamma}$), which in turn creates a lower or higher resistance to transport, depending on the effect of $\dot{\gamma}$ on the conformation. η_E may also be shear-thinning or shear-thickening. Shear-thickening fluids present difficulty in passing through tubes, orifices, and nozzles.

Rheopectic (antithixotropic) fluids are shear-thickening fluids whose η_a increases with time under constant or low $\dot{\gamma}$. Rheopexy is a property of linear macromolecules—rarely of polysaccharides—that is readily observed in lyophobic systems.

Plastic flow (unrelated to pseudoplasticity) is a linear response to τ after a critical τ (the yield point τ_0) has been exceeded. A plastic fluid is synonymously called a Bingham body.

A high $\dot{\gamma}$ or τ may cause a uniaxial orientation of random-coil molecules in the direction of flow, resulting in temporary flow birefringence as the primary linear structure uncurls. In every unit volume, there is a τ_0 created by the volume–temperature–pressure interplay expressed in Eq. (3.1). It is the uncurling resulting from $\dot{\gamma}$, τ, and T (Fig. 2) that causes streaming potential and lowers the magnitude of η and η_a.

η_i and η_a of ionic polysaccharides can be influenced by the order of addition of components, as illustrated in Fig. 3. η_i and η_a are maximum when the dispersed solute is allowed to hydrate fully before the electrolyte is added. An electroviscous effect is observed as an abnormally high η_i and η_a at very dilute concentrations (Fig. 4). As c_i increases, ionization is depressed and the electroviscous effect disappears; afterward, η_i vs c_i is that of neutral molecules (Pals and Hermans, 1952). Everett (1988) offered three reasons for electroviscosity—the distortion of the electric double layer during shear (the primary effect), double-layer repulsion between particles (the secondary effect), and a tertiary effect related to the hydrocolloidal diameter. The elimination of electroviscosity can reasonably be explained by curling, because the initial primary structure then becomes less expansive in an excess of the equivalent weight of counterions. Some polymers like CMC display an opposite electroviscous effect; they increase in η_{sp}/c as NaCl is

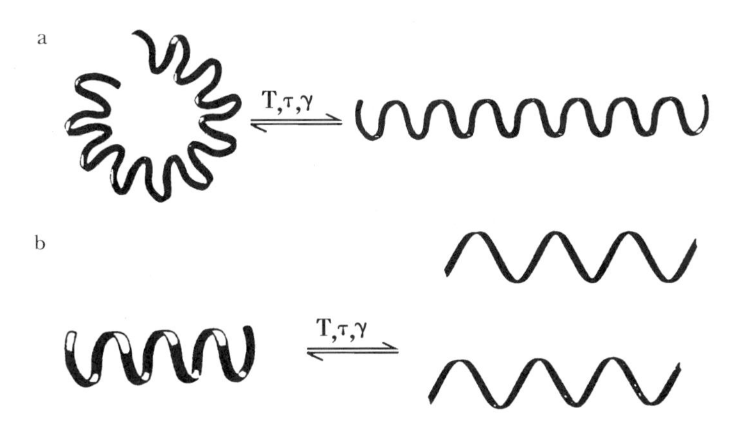

Figure 2 Illustration of the defibrillation of a polysaccharide double helix as a function of temperature (T) in a unit volume of solvent (water), flowing under shear rate (γ) and pressure (τ). Elongation of the single helices exposes a smaller cross-sectional area, resulting in birefringence and a lower circumferential resistance to flow (lower η). As a result of defibrillation (e.g., doubling of the microfibrils), number-average but not weight-average properties increase.

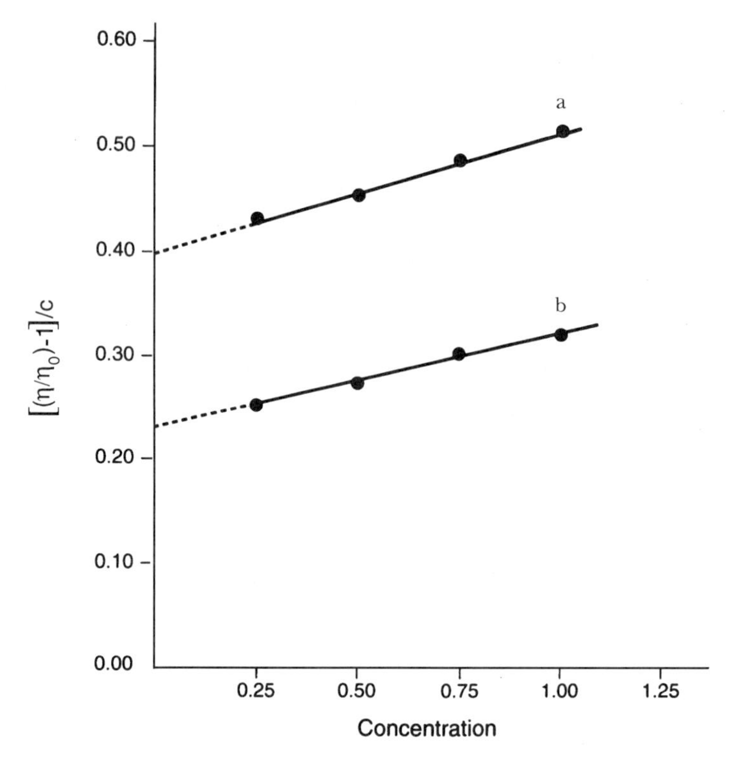

Figure 3 Viscosity profile of a 0.05% CMC sol, showing the effect on viscosity of order of addition of tartaric acid (TA) to water (a) before and (b) after dispersion of CMC.

added initially and then gradually decrease upon further additions (Dautzenberg *et al.*, 1994).

A viscoelastic fluid has the appearance of a solid body; it deforms and wholly recovers below τ_0, and only partly recovers above τ_0. From 0 to τ_0, the fluid undergoes an elastic conformational transition; above τ_0, the fluid undergoes an irreversible transition, whence the mass begins to flow toward a new equilibrium position. Carrageenan–water–polyol systems have been suggested to be industrially useful in consideration of their significant τ_0 followed by shear-thinning (Tye, 1988).

Viscoelasticity has advantages: for example, when butter and fruit jellies are spread over toast, they remain in place after the spreading force has been withdrawn. Paint is easily spread over a surface by the streaking of a paint brush, and remains in place when the streaking ceases.

Strain hardening is an abrupt, positive deviation of η_E from the Trouton rule as ϵ increases with time. In the view of Hwang and Kokini (1991), with reference to polysaccharides, this phenomenon is due to branch points acting as hooks, thereby increasing the resistance to flow.

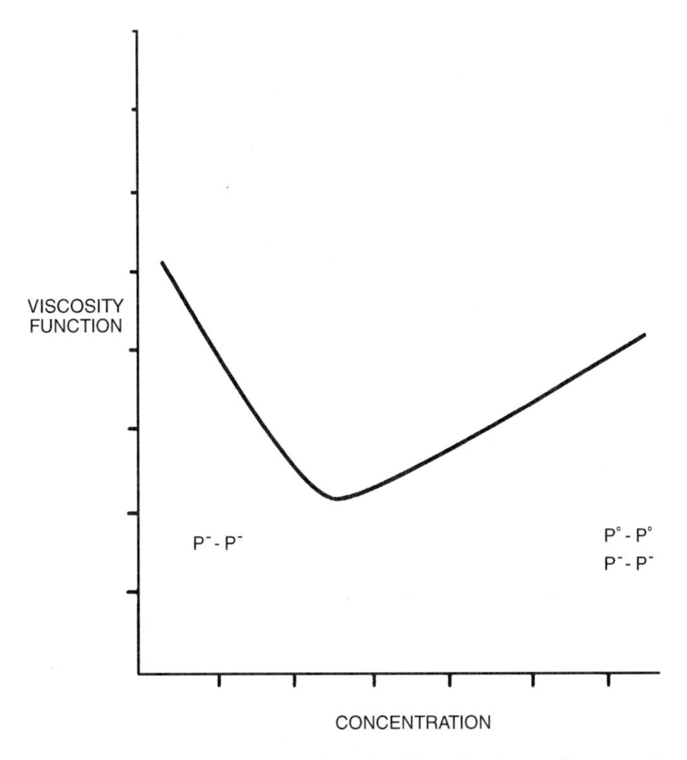

Figure 4 Typical viscosity response of a polysaccharide polyanion and a neutral molecule to concentration, showing electroviscosity in a dilute dispersion of the polyanion (negative slope segment) and linearity resulting from interactions and cancellation of electroviscosity (positive slope). P^- represents the polyanion and P° represents its neutral counterpart.

X. Variable-Path Processes

Path-dependent properties are a function of the sequencing of the transition steps from an initial to a final state, and it is only when each event is enacted in an identical way that the integrals of path-dependent functions are themselves identical.

Different polysaccharide variable-path processes were observed in a pectin–sugar–water–acid mixture dispersed at 105°C then cooled to 25°C, and a mixture dispersed at 50°C then cooled to 25°C. The higher-temperature gel was relatively stable, but the lower-temperature gel was unstable (Walter and Sherman, 1986). Heated agar sols gel when cooled to approximately 30°C, and they remain dimensionally stable to a reheating temperature of 85°C due to relatively permanent physical crosslinks below T_m (Lips *et al.*, 1988). The reheated gels follow a hysteretic pathway to melting at the higher temperature. Evaporation of a dilute polysaccharide sol to a film results in smaller pores in the film than if the pores were created by hydration of a xerogel. This sorption hysteresis results from the high volume of evaporating water that initially leaves large pores, until they contract in the xerogel, as its pore diameters attempt to narrow. In the reverse process, pores in a xerogel, made small by contraction, expand in size to accommodate the large volume of hydration water. Most polysaccharide xerogels are amorphous and consequently easily rehydratable.

A. Sols, Gels, and Pastes

All polysaccharides gel, albeit under different conditions. The designation of a polysaccharide as gelling or nongelling is merely a reflection of the ability of its aqueous dispersion to solidify under the prevailing conditions of food processing and preparation. Most food gels are made by heating and cooling a sol. Curdlan and konjac gums gel by neutralization of their alkaline solutions (Kanzawa *et al.*, 1989). Conformational transitions invariably accompany solation–gelation (Clark and Ross-Murphy, 1987; Rees *et al.*, 1982). The polysaccharide concentration in food gels seldom amounts to more than fractions of one percent, but enough solute–solute interactions initiate contact sites or junction zones that effectively crosslink such a small amount into an infinite (Flory, 1953) cooperative (Jeffrey and Lewis, 1978) network. The junction zones are nonequilibrium assemblies (Rees, 1969), dynamic and finite in size (Oakenfull, 1991) and constructed with lengths of about 600 atoms (Sperling, 1986) into localized regions of order whose lifetimes are transient around a constant mean. It is the instantaneous averaging of these lifetimes that gives a polysaccharide gel a macroscopic picture of uniformity and permanence. The uneven distribution of junction zones leads to crosslinking inefficiencies (Kulicke and Nottelmann, 1989) that in turn lead to structural inhomogeneities and uneven texture. The chain flexibility of random coils facilitates entanglement, crosslinking, and junction-zone formation. Microcrystalline regions are also sites of junction-zone formation, but by a different mechanism (parallelism). Starch gels are occasionally composites of gelatinized granules embedded in a crystalline amylose matrix (Miles *et al.*, 1985). In excess concentration, solute adds to an elementary structure (Dea *et al.*, 1972) until the first floc grows into a fractal aggregate confined only by the container's dimensions. Fractal aggregates are quaternary assemblies.

Kulicke and Nottelmann (1989) divided gels into three generic classes, viz., physical, ionotropic, and covalent. Physical gels are held together by hydrogen bonds and molecular entanglements; they expand when hydrated and contract when dehydrated: to Silberberg (1989) they are swollen mechanical and osmotic systems possessing the cohesive properties of solids in which a balance is struck between expansion and dissolution. Tanaka (1981) views them as fluid systems given form and maintained by a network of polymer strands under the net influence of attractive and repulsive forces. The cluster theory of gelation suggests that colloidal particles "stick" to each other almost irreversibly on contact, and form various spherical, needlelike or platelike shapes when the thermal energy, and hence Brownian motion, fall below the energy of attraction (Birdi, 1993). Walstra *et al.* (1991) postulated that some gels result from an enthalpy-driven aggregation mechanism of small particles with a high ratio of length to thickness. Michel *et al.* (1984) discounted the junction-zone mechanism of gelation for high-methoxyl gels in favor of an aggregation mechanism, but not necessarily exclusive of the junction-zone mechanism. Leloup *et al.* (1992) proposed an infinite network model for starch in which amylose combines with amorphous regions through an intermediate transition zone; in this model, the amorphous phase of dangling chains is responsible for the hydrodynamic behavior and porosity of the gel.

Physical hydrosols and hydrogels are theoretically interconvertible, with the possible exception of high-methoxyl-pectin hydrogels that do not normally revert to a sol by reheating (Walter and Sherman, 1986): the pectin contained therein was recovered by dialysis and was comparable (by η measurements) to the pectin before jelly formation.

Ionotropic gels develop from ionizable polysaccharides and are consequently pH- and electrolyte-sensitive; their water-absorption capacity rises with increasing concentration of ionic groups (Prud'homme *et al.*, 1989). Gelation is induced by low pH, and the gel strength arises from autocatalytic, cooperative bonding through ion mediation, whereby Ca^{2+}, for example, forms the first $R_nCOO{-}Ca^{2+}{-}OOCR_n$ bridge, followed sequentially by a series of like bridges between pairs of R_nCOO- (Grant *et al.*, 1973). Cooperative bonding supplements hydrogen bonds and entanglements in building gel strength and texture. A strong firm gel has a high incidence of junction zones, crosslinks, microcrystalline sites (Stipanovic and Giammatteo, 1989), and autocatalytic cooperative bonds.

Ionotropic gels have been made with alginate and Ca^{2+} in a cold process involving a stream of sodium alginate injected into a bath of calcium chloride (Kelco, 1986); the rate of delivery of the sol affects the final gel texture. In another process, a sparingly soluble calcium salt was dissolved or suspended in a mixed dispersion of alginate and pectinate, and Ca^{2+} was slowly generated by a lactone (Morris and Chilvers, 1984). Ionotropic polysaccharides gel also by monovalent cations, but the mechanism, presently incompletely understood, is different from autocatalytic cooperative bonding with divalent ions: the dehydration of carboxyl groups is believed to be involved somehow, and K^+ and Na^+ appear to behave differently in structure ordering and disordering (Miyoshi *et al.*, 1994).

Ionotropic gels are more acid-stable than the sols (Guiseley *et al.*, 1980) because of protonation of the electrolyte-sensitive acidic groups and immobilization of the molecules in the network.

Covalent gels develop from copolymerization with bifunctional crosslinkers: these are industrial-purpose gels having little or no relevance to food, except as aids in processing and research.

Chandrasekaran *et al.* (1988a) proposed Rees' mechanism (Rees, 1969) that random coils above T_m form double-helical junction zones upon cooling, which then aggregate prior to gelation. By themselves, the double helices do not cause gelation, but coordination complexes are required to be built with hydrated mono- or divalent cations (Chandrasekaran *et al.*, 1988c). Proof that double-helical junction zones alone did not cause gelation lay in the fact that ι-carrageenan in the presence of Li^+ formed double helices but did not aggregate and hence did not gel (Morris *et al.*, 1980).

Kanzawa *et al.* (1989) found that the width of microfibrils (50–250 Å) had an influence on gel structure, in contrast with viscous, nongelling dispersions in which shorter dimensions (10–20 Å) predominated.

Beet pectin and wheat-flour pentosans undergo an oxidative gelation (Neukom, 1976; Neukom and Markwalder, 1978) instigated by ferulic acid peroxidase through diferulic acid crosslinking. Crowe (1989) and Thibault *et al.* (1991) effected the same crosslinking with persulfate and chlorine, respectively.

Polysaccharide pastes are concentrated dispersions and suspensions with vastly diminished Brownian activity; they differ from gels by failing to undergo the liquid–solid (coil–helix) transition.

B. Emulsions and Foams

Polysaccharides may exercise a protective action in an emulsion and foam as a thin film at liquid–liquid (emulsion) and liquid–air (foam) interfaces. The hydrophile–lipophile balance in the macromolecules as well as ϕ_i determines whether or not the emulsion is an oil-in-water or water-in-oil dispersion (Vold and Vold, 1983; Dickinson, 1992).

C. Xerogels and Films

The anhydrous solid phase remaining after an apparently complete evaporation of water from a polysaccharide hydrosol is a polysaccharide xerogel, capable of retaining 20–30% water, yet appearing to be dry. Difficulty in removing residual water from polysaccharide xerogels makes a condition of 0% water virtually impossible. Dilute electrolytes initially present in the sol become concentrated as a result of the evaporation of water, leading to localized regions of high i with ramifications for the electrokinetics of the xerogel. Many young fruits and vegetables are polysaccharide hydrosols, but convert to xerogels during senescence.

Upon rehydration, a xerogel is capable of swelling to 10–100 times its dry volume without disintegrating. Such inordinate volume expansion is the reason why dehydrated fruit and vegetable food preparations require so much more liquid than their weight would otherwise suggest.

Films are two-dimensional xerogels. Composite polysaccharide films contain complementary cosolutes, each designed to fulfil one or more shortcomings of the other(s). Polysaccharide films have a low permeability to oxygen, a high permeability to moisture, and low tensile strength. The permeability is sensitive to the number and distribution of segments (Silberberg, 1992), i.e., to the number and distribution of junction zones, because these block the solvent's path. In other words, permeability is highest where junction zones are least dense. According to Kester and Fennema (1986), a high-moisture gelatinous polysaccharide acts more as a short-life ''sacrificing agent'' than as a moisture barrier, inasmuch as there is a preferential release of moisture from the moisture-laden coating to that of the packaged item. Films with triple helices are apparently stronger than films without (Schulz *et al.*, 1992).

Polysaccharide films are used to protect the delicate flavor and aroma of fruits and vegetables from a deleterious physical environment. Cellulose xerogels do not readily rehydrate because of the strong hydrogen bonding of the microfibrils in the dry state, but they make excellent protective membranes and coatings, relying on the tensile and compressive strengths of their interwoven microfibrils and on plasticizers (e.g., glycerol and dextrin) to overcome their characteristic brittleness and hardness.

Non-free-draining water has a minor plasticizing effect on polysaccharide films. Lipid plasticizers increase hydrophobicity and decrease the rate of moisture transport across the film. Thermosetting cellulose films obstruct lipid migration into frying foods (Nelson and Fennema, 1991). Engineering cellulose films for specialized properties (high solute retention, high solvent exchange, etc.) requires control of a number of variables (Shen and Cabasso, 1982).

D. Aerosols

An aerosol is a dispersion of discrete particles in a stream of gas. Starch and cellulose aerosols are potential fire hazards in granaries where friction between the moving, micronized particles causes electrification, whereupon separate accumulations of positive and negative charges may discharge as an electric spark and ignite the combustible solute (contact electrification; synonymous with triboelectrification; Ross and Morrison, 1988).

E. Suspensions

Suspensions are macroscopic, heterogeneous, usually liquid–solid systems that contain all or a fraction of solute larger than colloidal dimensions. A polysaccharide suspension may consist of a continuous network of irregularly shaped particles (Walstra *et al.*, 1991). In these multicomponent systems, particles dissimilar in size diffuse at dissimilar rates. The large fraction refracts and the colloidal fraction scatters light. In number- and weight-average measurements, the multicomponent phase properties are indistinguishable from those of a homogeneous phase, but they have much larger variances. Examples of polysaccharide food suspensions are tomato ketchup, unclarified fruit juices and beverages, fruit pulp, and chocolate milk.

XI. Stability and Instability

To the food processor and preparer, a stable polysaccharide dispersion is one that has a long shelf-life. For maximum utilitarian benefit, most dispersions are maintained in a kinetically stable state instead of a thermodynamically stable state. Absolute thermodynamic equilibrium is the existence of the solute in one spherical mass at the bottom of a container where each component of a dispersion reverts to its ground-state energy level. Thermodynamic stability is thus antithetical to superior food quality: it is mostly observed between macromolecules with a different chemical composition and those with the capacity to form only weak bonds with each other (Tolstoguzov, 1993). Kinetic stability requires the particle phase to remain dispersed in prolonged metastable equilibrium under a prescribed set of conditions. By thixotropy, the life of a suspension of discrete, heterogeneously shaped particles in an aqueous medium is prolonged over that of homogeneously shaped particles, because the semipermanent structure they create takes a longer time to dehydrate and phase-separate spontaneously before settling. Destabilization is the process of a dispersion's progression from kinetic to thermodynamic equilibrium (see Chapter 7, Fig. 5).

Myers (1960) recognized two kinds of stability, viz., inherent stability as a function of time, and induced stability as a function of stimuli. Induced kinetic stability necessitates the expenditure of work exceeding ξ.[12] Acceptance of a stabilization role for water has not been universal; Jirgensons (1946) argued against it, enunciating instead the influence of particle shape and the mutual chemical affinity between atomic groups at the solid–liquid interface.

12. The unit of a potential is the joule (force times distance per unit charge).

By the logic of the capacitor model, "salting in" is a stabilizing mechanism whereby ζ is lowered and in turn Q/ζ is elevated. The capacitor model does not explain "salting out" as easily as does the electrolyte effect of i directly on the Debye length.

Steric stabilization differs from electrostatic stabilization in not being a function of a net force, but of the thickness of an adsorbed layer. When ϕ_i equals 5–10%, stabilizing and destabilizing forces extend beyond the length of the electrostatic, interparticle barrier (Cabane *et al.*, 1989). At this distance, attraction and repulsion are inconsequential, and electrolytes therefore have little effect. Bergenstahl (1988) proposed that the steric stabilization of emulsions by gums in the presence of a surfactant involves adsorption of the gum on the surfactant to form a combined structure constituted by a primary surfactant layer covered by an adsorbed polymer layer.

Destabilization is signalled by incipient flocculation. The latter occurrence was considered by Vold and Vold (1983) to be the initial reversible stage of aggregation. Coalescence and coagulation are qualitatively synonymous terms. Hiemenz (1986) made a distinction by considering flocculation as a process that allows small particles to retain their identity but lose their kinetic independence, and by considering coalescence as a loss of particle identity in favor of larger particles. A floc is less dense, because it occludes more water than a coagulum. Gelation, precipitation, and crystallization are stages of one continuum in a sol between dispersion and deposition.

Small quantities of polysaccharides can flocculate a dispersed phase through bridging (Ward-Smith, *et al.*, 1994), whereby one attached molecule with other adsorption sites along it may attach itself to another or more surfaces, acting as the "bridge"; this phenomenon is called bridging flocculation. A bridge may instead cause steric stabilization of the dispersed phase. In the view of van Oss (1991), steric stabilization is predominantly a polar repulsion between macromolecules that is influenced not by Brownian activity, but by osmosis.

Depletion flocculation arises when a large unadsorbed, flocculating cosolute molecule does not fit properly into a small interparticle volume at the interface and the cosolute molecule accompanied by solvent is consequently expelled from the interface. As a result, the interparticle distance is shortened, causing an approach to x_e and flocculation. Depletion stabilization is possible if the particle–cosolute attraction is greater than the particle–particle or cosolute–cosolute attraction.

Cations flocculate hydrosols at critical ionic strengths that vary with polymer concentration, particle size, temperature, etc. For mono-, di-, and tervalent cations, in ascending order of strength, the critical concentration is 1, 0.03, and 0.001 mmol/L (Vold and Vold, 1983). In the presence of flocculating cations, during the progression from kinetic to thermodynamic stability, dispersed polysaccharides gradually lose mobility ($-\Delta S$), their surfaces merge, the particles grow larger and are fewer.

TABLE II
Conditions and Mechanisms of Stability of Polysaccharide Sols

Condition	Mechanism
Micronization (Homogenization[a])	Buoyancy Brownian motion
Hydration	Capacitance Maximum excluded volume Minimum excluded volume effects Coil volume expansion
Dilution	Capacitance Maximum excluded volume Minimum excluded volume Coil volume expansion Salting in
Heating	Brownian motion $(+\Delta H, +\Delta S)$
Low acidity	Dissociation Electroviscosity ζ potential
Neutralization	Soluble salt formation
Cosolute addition	Protective colloid action Steric stabilization
Viscosity increase	Slow sedimentation rate

[a] In this context, homogenization refers to reduction of particle size by passage of the disperse system through a small aperture under pressure.

Undesirable emulsions and foams in food-processing operations are "broken" by antifoaming agents whose exact mechanisms of action are uncertain, although macromolecules performing this function are known to create ordered assemblies at the interface and are themselves excellent emulsifiers. Antifoaming agents thin and weaken small regions of an adsorbed film (Shaw, 1992); all rapidly lessen $\sigma_{o,i}$ to the extent that the attractive forces between the antifoam and one of the phases (adhesion) exceed the attractive forces between like molecules (cohesion).

Creaming is the opposite of sedimentation; it occurs when the solute phase (usually oil) has a density less than that of water.

The conditions and mode of action that contribute to the kinetic stability of polysaccharide sols are listed in Table II.

A. Aging and Phase Separation

In polysaccharide dispersions, a constant-temperature, constant-pressure separation of any or all of the solute from its dispersion medium may be effected spontaneously by time (aging) or may be actuated by external stimuli (induction). Concomitantly, viscosity parameters R_g and ϕ are lowered. During the process, smaller-size solute particles tend to grow into larger-size solute particles—a phenomenon termed Ostwald ripening. High DP, solute and electrolyte concentrations, low temperatures, nonsolvents, evaporation, storage-temperature fluctuations, and stimuli as innocuous and unobserved as vibrations in the surroundings accelerate either process. The purpose of a protective colloid is to extend the duration (shelf-life) of the apparent monophase. In either spontaneous or induced destabilization, the solute is transformed from its initially high ΔE to a final low ΔE state.

B. Coacervation

Coacervation is the separation into two liquid phases of a ternary dispersion, each phase containing a preponderance of one solute and a minor concentration of the other, and vice versa: each phase is a coacervate. The event is simple coacervation if the cosolutes have identical charge; it is complex coacervation if the cosolutes are oppositely charged (Jirgensons and Straumanis, 1962). Either phase may develop a network independently of the other (Moritaka *et al.*, 1980), or one phase may be suspended as droplets in the other. Alternatively, one solvent-depleted phase may contain the two cosolutes, while the other phase is preponderantly solvent.

C. Syneresis

Syneresis is the tendency of gels to release spontaneously small volumes of liquid, occasioned by the rupture of weak bonds, under an internal τ. This mechanism lowers $+\Delta E$ (Walter, 1991), as the gel attempts to return its components to their respective ground states. Rigid gels are prone to synerize, because the elastic component does not possess the mobility of the viscous component and, consequently, phase separation is the only energy-releasing alternative to viscous flow. Soft gels synerize when τ exceeds the bonding strength. Syneresis is sometimes accelerated by freeze–thaw cycles.

Starch and pectin gels are noted for their ability to synerize; xanthan has received wide acclaim as a syneresis-controlling polysaccharide (Rocks, 1971); κ-carrageenan gels are firm, brittle, and given to syneresis, whereas ι-carrageenan gels are soft, elastic, and syneresis-free (Roesen, 1992). Corn starch all but eliminated syneresis in a 4% curdlan gel that had been subjected to freezing and thawing (Nakao *et al.*, 1991); one possible explanation of the amelioration is that corn starch is a humectant. Other syneresis-controlling practices include pH and soluble–solids adjustment (Konno *et al.*, 1979; Hercules Inc., 1985). In pectin jellies in which the defect is frequently observed, syneresis may be avoided by simply changing the sugar concentration or the pectin (Hercules Inc., 1985).

D. Sedimentation

A sediment is a solid phase separated from its dispersion medium in a relatively solvent-free condition: the process is called sedimentation or deposition. The rate of sedimentation depends on η_o, ϕ_i, $\overline{M}_i$, particle size, and the density difference between the solvent and solute (Scholte, 1975; Harding *et al.*, 1991a). The density of highly hydrated particles is approximately equal to the density of water: a large volume of non-free-draining water may therefore cause a floc to remain suspended almost indefinitely. Very small density differences do not provide enough of a gradient to affect rapid deposition.

A polyanion's sediment layer is more diffuse than that of a neutral polysaccharide, because of interparticle carboxyl-charge repulsion. Easy repeptization of uronan-containing sediment in juices and wines presents difficulty during filtration and decantation.

E. Encapsulation

Polymers may be induced to encapsulate other molecules by a variety of means (Risch and Reineccius, 1995) as diverse as dipping, spray-drying, extrusion, evaporation, and coacervation: each technique has its special applications, strengths, and weaknesses. Advantages in common are the protection and slow release of the encapsulate. In any of the mechanisms, a coagulable polymer precipitates around a core of labile material. Polysaccharides are regular encapsulating polymers (Risch and Reineccius, 1995); acacia gum is particularly efficacious because of its protein content.

In a microencapsulation method, the encapsulate—usually an oil, flavor, enzyme, or medicinal—is emulsified in a dilute aqueous gelatin sol, a polysaccharide is added, and conditions are adjusted to favor coacervation. The encapsulate should not be truly soluble in the solvent or the cosolutes and the cosolutes should be differentially soluble in the liquid solvent. As much as 60–98% of the labile substance may be harvested by microencapsulation to yield microcapsules in the form of a free-flowing powder (Sirine, 1968).

In a spray-drying method of encapsulation, Zhao and Whistler (1994) suspended starch to a concentration of 30% in water containing 0.1–1.0% gelatin or any of a number of polysaccharide bonding agents: the suspension was forced through an orifice (2 mm diameter) under 80–100 psig at an inlet temperature of 120°C and outlet temperature of 76°C. Porous 10–40-nm-diameter spherical capsules were obtained that were then immersed in peppermint oil. After diffusion of the peppermint oil into the capsules, the spheres were rinsed free of oil and coated in a fluidized bed with a 3%

dispersion of the bonding polysaccharide. The finished capsules contained 33–48% peppermint oil.

What Tye (1988) called "entrapping technology" involves dropwise gelation in a KCl bath of an emulsified solute and carrageenan (1% in distilled water). The dried (and presumably washed) gel capsules were reportedly capable of retaining in the carrageenan network any dissolved or emulsified cosolute.

XII. Summary

Macromolecular conformations and reversible order–disorder and disorder–order transitions are highly sensitive to solvent, temperature, pressure, pH, water activity, and metal ions. Polyanions are distinguished from neutral molecules by their sensitivity to electrolytes. Whereas synthetic polymers do not normally dissolve or disperse spontaneously, some polysaccharides may do so in water (hydration), given their strong hydrophilicity.

The random coil is a high-energy conformation and the helix is a relatively low-energy conformation: the former is disordered and the latter is ordered. The coil–helix transition is consequently attended with a large negative entropy change and a larger negative enthalpy change. The energy status of a final product depends on the treatments it underwent during the final stages of processing and preparation. High-energy food dispersions require special treatment for prolongation of their shelf-life: examples of such treatments are the inclusion of a protective colloid and maintaining high solvent viscosity.

Polysaccharide dispersions phase-separate spontaneously, a phenomenon called aging. Phase-separation may be induced in special systems, under controlled conditions (e.g., encapsulation), to industrial and commercial advantage.

CHAPTER 4

Concentration Regimes and Mathematical Modeling

I. Introduction

Polysaccharide uses span dilute, semidilute, and concentrated regimes, from fractions of a percent in a hydrosol to higher than 90% in a xerogel. At the lower limit, solute–solute interactions are minimum if not entirely eliminated, and a macroscopic property (P) is theoretically the sum of the properties of independently acting molecules. At the upper limit, P is the combined property of multiples of molecules seen as clusters, each cluster acting as a hydrodynamic unit. A critical micelle concentration (c^*, Fig. 1) differentiates the additive response of single molecules from that of clusters. When the number (n_i) of equivalent spheres each of weight w_i increases in a volume of water (V_o), the solution density $n_i w_i / V_i$ increases until the minimum x_e is reached when the dispersion medium becomes saturated. c^* depends intensively on the DP and extensively on $n_i w_i$, under the joint influence of conformation, cosolutes, nonsolvents, T, $\dot{\gamma}$, and τ. High $\overline{M}$ polysaccharides are less energetic and have a lower c^* than do low $\overline{M}$ polysaccharides, causing the molecules to cluster or precipitate sooner.

II. Concentration Regimes

There are four to six concentration regimes (Dautzenberg *et al.*, 1994). Edwards (1966) classified them into three broad types on the basis of the number of polymer chains present, the number and length of the monomer constituting them, and the volume of the monomer relative to the volume V_i. At use levels as low as parts per thousand in food, no one definition of "dilute," "semidilute," or "concentrated" encompasses the range of weight–volume or volume–volume concentrations necessary to elicit a par-

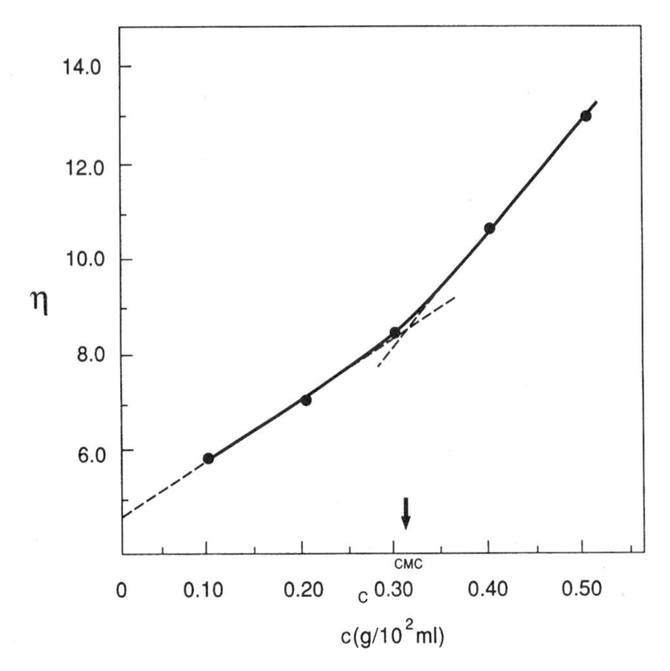

Figure 1 Viscosity–concentration (η vs c) profile at 26°C of an aqueous pectin dispersion showing the critical micelle concentration (c^*).

ticular physical response or human sensation: for example, a constant weight each of high-methoxyl and low-methoxyl pectin with the same DP gels under different conditions involving water, sugar, acid, calcium, and heat. Without acid, a heated high-methoxyl-pectin jelly formula remains a sol, but with acid, it transforms to a gel. Gellan gum, requiring no sugar or acid, gels at 0.05% concentration in water, whereas 0.5–0.8% pectin requires sugar and acid. Similarly, pectic acid, an ionic polysaccharide, and dextrin, a neutral polysaccharide, make very thin fluids at 2% concentration, but CMC and guar gum, also an ionic and a neutral polysaccharide, respectively, make extremely viscous fluids at one-tenth that amount. The singular event common to the three gels is a high incidence of solute–solute contacts.

A. The Dilute Regime

Doi and Edwards (1986) characterized a dilute solution as one of sufficiently low concentration that the polymer molecules are separated from each other. Dilute sols are normally characterized by a linear dependence of P on c_i and often by Newtonian flow. A constant rate of change of η_i vs c_i is desirable for uniform fluid texture in liquid foods like dietetic beverages and clarified fruit juices, and for uniform delivery in flavor-release capsules (Baines and Morris, 1988) where slow release is essential. Linearity holds for neutral polysaccharides in water and for ionic polysaccharides in an excess of electrolytes.

Dilute dispersions have a very small ϕ_i and, especially in the case of polyelectrolytes, a dielectric constant that is unfavorable for cluster formation (Eisenberg and King, 1977). Dilution may not eliminate the structuring effected by polysaccharides in water, because structuring does indeed exist in "thin" liquids (Schenz and Fugitt, 1992).

B. The Concentrated Regime

At the upper limit of the c_i range, v_f decreases to a minimum as the molecules are progressively immobilized, effectively making "good" and "poor" solvents functionally indistinguishable. In this regime, viscosity merges into elasticity, P becomes independent of c_i, and the dispersion simulates the behavior of a molten polymer.

C. The Semidilute Regime

The vicinity of c^* is transitional to the dilute and concentrated regimes across a narrow c_i range where the topologically linear but conformationally distorted coils touch, overlap, and entangle, to the extent that free movement of segments is restricted and diffusion is unidirectional (reptation), because of the constraints on radial motion imposed by neighboring segments. Reptation was theorized by de Gennes (1979) and experimentally proven by Russell *et al.* (1993). Cluster formation and complex flow originate in entanglements in this region. A careful study of η_i vs c_i (for some polysaccharides) reveals an inflection closely preceding c^* where, it is claimed, molecules actually begin to entangle (Launay *et al.*, 1986). In the transition zone of c^*, P is scaled to a power function of c/c^* (Dautzenberg *et al.*, 1994). The contact points are dynamic and identical, except that they have much shorter lifetimes below c^* than above. Reversible gelation is ascribed to few entanglements of short-life contact points with insufficient combined strength and duration to maintain a suprastructure indefinitely, unlike those in an irreversible gel. Numerous contact points encumber diffusion.

Random-coil polysaccharides make strong films, because they are given to a high incidence of long-life contact points; for the same reason, they are good carriers of flavor. Notably, concentration had no effect on flavor release from a nongelling xanthan dispersion (Baines and Morris, 1988).

Time-dependent flow and viscoelasticity begin to be evidenced in the semidilute regime where ϕ_i relative to ϕ_i in the dilute regime is high, the dispersion remains sensitive to T, and P is equally a function of micellization as of colligative action. Pseudoplasticity was observed in 0.5–2.0% guar gum and 0.5–1.2% xanthan gum; thixotropy was observed in 0.8–1.2% furcellaran (Rao and Kenny, 1975). Okiyama *et al.* (1993) observed dilatancy at a low $\dot{\gamma}$ in an aqueous 0.7% dispersion of a disintegrated bacterial polysaccharide. Some gums (e.g., CMC and guar) change from dilatant to plastic flow (Balmaceda *et al.*, 1973). Thixotropic fluids are easily mistaken for pastes and gels, due to their self-assembly into suprastructures at rest.

The dimensionless product $c[\eta]$ is defined as the coil overlap parameter; it provides information about the changing nature of the interactions in a dispersion (Blanshard and Mitchell, 1979; Morris *et al.*, 1981). For dilute dispersions, i.e., below c^*, the slope of $\log(\eta_{sp}/c_i)$ vs $\log(c[\eta])$ universally approximates 1.4. At the upper practical extreme, with exceptions (especially the galactomannans; Morris *et al.*, 1981), the slope increases sharply to 3.3, illustrating wide deviations from Newtonian flow in the segment approaching elasticity. The deviations are significant when $5 \leq c_i[\eta] \leq 10$ (Barnes *et al.*, 1989).

III. Mathematical Modeling

P is characterized mathematically as a scalar variable (ν) raised to the magnitude of one or more systemic constants. The defining equation may be a linear ($P = m\nu + \rho$), a quadratic ($P = j\nu^2 + j'\nu + j''$), an exponential ($P = \kappa e^{\nu} + \kappa'$), or a power ($P = k\nu^{\alpha} + \omega$) function of ν, where m, ρ, j, j', j'', ω, κ, κ' and a are algebraic constants. Assisted by these models, the concept of the equivalent hydrodynamic sphere, under θ conditions, facilitates characterizations of properties and calculations of various size parameters. Many aqueous polysaccharide properties are a function of M to an asymptotic limit. Linear flow is readily amenable to mathematical analysis and, not surprisingly, a number of equations to measure η have thus been derived. The Maxwell and Voigt–Kelvin models (Kaelble, 1971; V. N. M. Rao, 1992) combine mechanics and mathematics to demonstrate that viscoelasticity and elasticity are mechanisms of storage and dissipation of ΔE. Although η is a simple property to measure, rheological data can nevertheless be difficult to interpret (Barker and Grimson, 1991). In contemporary methodologies, size- and shape-related complexities have largely been minimized, where possible, by extrapolation to infinite dilution ($c_i = 0$).

A. The Stokes Equation

In 1851, Stokes derived Eq. (4.1) from the model of solid spherical particles falling independently through a homogeneous liquid without Brownian motion, slippage, and wall effects. Slippage is an inconstant rate of fall; wall effects refer to axial orientation in the outermost planes of fluid in contact with a surface, and the differential velocity of flow in the outermost and innermost planes of a fluid in a confining tube:

$$\partial x/\partial t = k(\mathbf{d_s} - \mathbf{d_o})gr_i^2/\eta_0 . \tag{4.1}$$

The spheres with radius r_i and density $\mathbf{d_s}$ fall through a solvent with density $\mathbf{d_o}$ and viscosity η_o at a rate $\partial x/\partial t$; **g** is gravity. F, counterbalancing gravity, is equal to the product of f_c and $\partial x/\partial t$, i.e.,

$$F = f_c(\partial x/\partial t). \tag{4.2}$$

The variables in Eqs. (4.1) and (4.2) are conventionally expressed in cgs[13] units. For a spherical geometry (Hiemenz, 1986),

$$f_c = 6\,\mathrm{pi}\,\eta_o r. \tag{4.3}$$

Equations (4.2) and (4.3) show that molecularly homogeneous solute providing a larger f_c settles more slowly than does solute providing smaller f_c. Alternatively stated, larger particles settle more slowly than smaller particles with the same density, barring hydration. Other empirical offshoots from the Stokes law were attempted, but complications arose from an initial lack of awareness of the contributions of hydration to particle factors (Mehl *et al.*, 1940).

B. The Poiseuille Equation

V_i flowing under τ in t_i seconds through a cylindrical tube of radius r and length l is inversely proportional to η_i and the fourth power of r:

$$V_i = kr^4 t_i \tau/(\eta_i l). \tag{4.4}$$

The dimensions of η_i are mass per unit distance per unit time, which, in cgs terminology when τ is in dynes (g cm s^{-2}), converts to poise (g cm^{-1} s^{-1}) (Appendix 1). In mks[14] terminology, F is in newtons (kg m s^{-2}) and the η_i dimensions cancel to kilograms per meter per second (kg m^{-1} s^{-1}). One

13. Abbreviation for centimeter–gram–second.
14. Abbreviation for meter–kilogram–second.

dyne equals 10^{-5} newton (N). τ expressed in newtons per meter squared is a pascal unit (Pa); 1 Pa = 1 N m^{-2}. The SI unit of η is the pascal second (Pa s). A ratio of η in cgs units and η in mks units [(g cm^{-1} s^{-1})(kg m^{-1} s^{-1})$^{-1}$] cancels to 0.1; 1 poise = 0.1 Pa·s.

Equation (4.4) is an unwitting statement that the velocity (l/t) of a sol's planar flow is inversely proportional to η_i. A capillary viscometer is designed to maintain r, l, V, and τ (1 atm) constant, so that η_i is directly proportional to t_i. The generalized equation for a single measurement (single-point viscometry) is

$$\eta_i = kt_i . \tag{4.5}$$

The unit of capillary η is the stoke, defined as poise per density, which reduces to centimeters squared per second. Having no reference to mass and force, capillary η is also referred to as kinematic η. The $\eta_i - c_i - t_i$ relationships have been formalized into a series of η functions (Table I).

If t_o and t_i are measured for a series of dispersions at different c_i and the data are plotted as η_{sp}/c_i vs c_i and extrapolated to $c_i = 0$, the intercept is $[\eta]$ in a volume per unit weight. $[\eta]$ reflects the magnitude of the interactions between flexible molecules in energetically favorable and unfavorable solvents (Alfrey *et al.*, 1942; Alfrey, 1947; Berth *et al.*, 1982); it is strongly influenced by T and i_s. The slope of η_{sp}/c_i vs c_i in a unit of volume squared per unit weight squared has functional significance; the steeper it is, the more thickening power component i has. For convenience, the absolute percentage value of c_i may hereinafter be used when a numerical remainder

TABLE I
Equations Relating a Dispersion's Flow Time (t_i) and Viscosity (η_i) to the Flow Time (t_o) and Viscosity (η_o) of the Dispersion Medium Containing Solute in Weight per Volume Concentrations (c_i) at a Constant Temperature

Viscosity parameter	Equation
Relative viscosity (η_{rel})	$\eta_i/\eta_o = t_i/t_o$
Specific viscosity (η_{sp})	$(\eta_i - \eta_o)/\eta_o = (t_i - t_o)/t_o$ $(\eta_i/\eta_o) - 1 = (t_i/t_o) - 1$
Viscosity number (η_{sp}/c_i) (Reduced specific viscosity)	$[(\eta_i/\eta_o) - 1]/c_i = [(t_i/t_o) - 1]/c_i$
Inherent viscosity	$\ln(\eta_i/\eta_o)/c_i = \ln(t_i/t_o)/c_i$
Intrinsic viscosity ($[\eta]$) (Limiting viscosity number)	$[(\eta_i/\eta_o) - 1]/c_i = [(t_i/t_o) - 1]/c_i$ $\lim c_i \to 0$ $[\ln(\eta_i/\eta_o)]/c_i = [\ln(t_i/t_o)]/c_i$ $\lim c_i \to 0$

of η_{sp}/c_i is given. Capillary η data are precise and linear, once t_i approximates 100 s, V/t approximates 0.1 cm^3 s^{-1} (Jirgensons and Straumanis, 1962) and $t_i/t_o < 2$ (Allcock and Lampe, 1981).

Measurements taken from a series of different c_i arrived at *in situ* by dilution in specially designed viscometers comprise the capillary viscometry technique known as dilution viscometry. Neither single-point nor dilution viscometry is suitable for suspensions, because of the unreliability of their t_i resulting from heterogeneities of particle size, shape, and interaction. Variations in t_i are conducive to slippage, wall effects, and turbulence.

A Reynolds number, computed as the dimensionless ratio of the length of the cylinder multiplied by the flow velocity and the kinematic η [$(cm^2\ s^{-1})(cm^2\ s^{-1})^{-1}$], is a ratio of inertial and viscous forces (Van de Ven, 1989). Newtonian flow occurs below a Reynolds number of 2200; turbulent flow occurs above 2200.

In a crude adaptation of viscometry known as Bostwick consistometry, t_i is held constant and l is measured as a function of η_a. This technique is suitable for liquid suspensions, e.g., tomato ketchup, that need not be subjected to rigorous quality control.

A different experimental design, called rotational viscometry, exploits the principle of fluid resistance, whereby concentrated dispersions and suspensions at and above c^* are sheared between two surfaces moving with different velocities relative to each other at constant or variable τ. Time dependence is measurable by rotational viscometry but not by capillary viscometry.

C. The Huggins Equation

Huggins (1942) derived Eq. (4.6) to characterize η_i vs c_i. The frictional coefficient (k) was included to account for "the sizes, shapes and cohesional properties of long-chain, neutral molecules":

$$\eta_{sp} = [\eta]c_i + k[\eta]^2 c_i^2, \tag{4.6}$$

$$\eta_{sp}/c_i = [\eta] + k[\eta]^2 c_i. \tag{4.7}$$

k, also called the Huggins interaction coefficient, is alleged to be specific to the particular solute–solvent system. Equation (4.7) with slope $k[\eta]^2$ is the linear form of Eq. (4.6). The steepest slope may sometimes betoken the poorest solvent, due to solute–solute interaction, and at other times betoken the best solvent, due to solute–solvent interaction (Alfrey, 1947). Substituting β for $k[\eta]^2$, the effect of c_i on η_{sp} is shown by the exponents in an expansion series [Eq. (4.8)] to exaggerate small property differences:

$$\eta_{sp}/c_i = [\eta] + \beta c_i + 2\beta c_i^2 + 6\beta c_i^3 + \cdots . \tag{4.8}$$

This exaggeration makes possible the characterization of polysaccharides by coordinate orientation (Walter, 1991). For routine purposes, accuracy to more than the squared term is seldom required.

By use of the Huggins interaction coefficient, a configurational distinction was made between guar gum and locust bean gum (Elfak *et al.*, 1977); with it, solute–solute interactions were indexed (Launay *et al.*, 1986). A significant increase in β was measured for different classes of galacturonans dispersed in water containing varying amounts of ethanol, when the ethanol concentration was 15–25%: there was no reliable trend in η or k', by itself, but the combined change (β) was unmistakable (Walter and Sherman, 1988). The increase was attributed to the higher frequency of solute–solute contacts in the increasingly hydrophobic medium (see Fig. 2 in Chapter 7). Aging lowers $[\eta]$ significantly (Walter and Sherman, 1983).

D. The Martin Equation

Equation (4.9) is a logarithmic equivalent of the Huggins equation that obscures the exaggerated effect of the exponent and yields a wider linear range:

$$\log(\eta_{sp}/c_i) = \log[\eta] + k[\eta]c_i . \tag{4.9}$$

E. The Kraemer Equation

The Kraemer equation is another logarithmic equivalent of the Huggins equation:

$$(\log \eta_{rel})/c_i = [\eta] + k[\eta]^2 c_i . \tag{4.10}$$

The Huggins and Kraemer lines plotted in the same graph converge at $[\eta]$.

F. The Schulz–Blaschke Equation

One form of the Schulz–Blaschke equation is stated as the second-order relationship

$$[\eta] = (\eta_{sp}/c_i)/(1 + k\eta_{sp}). \tag{4.11}$$

$(1 + k\eta_{sp})/\eta_{sp}$ is the slope of the locus delineated by $[\eta]^{-1}$ vs c_i (Appendix 2). This equation permits a wide range of linearity (Carpenter and Westerman, 1975) and is the most suitable for computing $[\eta]$ from a single concentration (Vink, 1954), although such computations are disfavored because they may conceal nonlinearity.

G. The Newton Equation

According to Newton's law [Eq. (4.12)], τ generated in a flowing dispersion over t is a function of η_i:

$$\tau = \eta_i \gamma / t = \eta_i \dot{\gamma}. \tag{4.12}$$

Newtonian fluids comply with Eq. (4.12). When $\eta_i \neq \tau/\dot{\gamma}$, η_i is replaced by η_a and $\dot{\gamma}$ must be stipulated. Plastic flow complies with Eq. (4.13) and is normally linear after τ_0:

$$\tau - \tau_0 \doteq \eta_i \dot{\gamma}. \tag{4.13}$$

The counterpart of $\dot{\gamma}$ in elastic fluids is the strain rate ($\dot{\epsilon}$).

H. The Power-Law Equation

The general equation for viscometry of non-Newtonian dispersions is

$$\tau = K\dot{\gamma}^{\nu}, \tag{4.14}$$

$\nu \neq 1$. Substituting for τ from Eq. (4.12), Eq. (4.14) becomes

$$\eta_a = k\dot{\gamma}^{(\nu - 1)}. \tag{4.15}$$

k is the consistency. Shear-thinning is characterized by $0 < \nu < 1$ and shear-thickening is characterized by $\nu > 1$. Launay *et al.* (1986) considered a major deficiency of Eq. (4.15) to be its prediction of infinite η, instead of Newtonian η, at $\dot{\gamma} = 0$. One advantage of Eq. (4.15) in rotational viscometry is that it enables the measurement of η_a and time constants of thick fluids.

Over an extended range of $\dot{\gamma}$, log η_a vs log $\dot{\gamma}$ reveals two segments, one at the lower and the other at the upper Newtonian region, where $\beta = 0$. η_a in each segment is referred to as zero-shear η. The rotational viscometry of polysaccharide systems begins with measurements at very low $\dot{\gamma}$ in the upper Newtonian region and passes through a shear-thinning interval to the lower Newtonian region.

I. Hooke's Equation

The basic equation for elasticity is Hooke's law, stated as

$$\tau = \mathbf{G}\epsilon. \tag{4.16}$$

G is the modulus of elasticity (synonymous with elastic modulus, rigidity modulus, bulk modulus, stress relaxation modulus), depending on the experimental design. **G** is remarkably sensitive to T; it increases from T_m to T_g, depending again on the degree of ionization (Brondsted and Kopecek, 1992), and decreases with nonsolvent and electrolyte additions to a critical swelling ratio, whence it increases again upon further additions (Oppermann, 1992). The larger the **G**, the firmer is the gel. The reciprocal of **G** (ϵ/τ) is the compliance.

By a process called creep, a viscoelastic fluid subjected to an instantaneously applied constant τ gradually deforms, storing energy $(+\Delta E)$ in the process. Upon the release of τ over a time interval (t), ϵ slowly returns to its original shape in an attempt to cancel $+\Delta E$: hence, creep is described as retarded elasticity. The recovery stage is called relaxation, which can occur over an indefinitely long interval; the longer the interval, the more elastic than viscous is a gel. In a creep test (Fig. 2a), τ is held constant until ϵ gradually increases to a maximum (ϵ_{max}) characterized by the cessation of flow $(d\epsilon_{max}/dt = 0)$. The creep curve demarcates three stages in a pressurized, viscoelastic fluid—an initial, τ-independent, completely recoverable (elastic) response, followed by variable-rate deformation (retardation) from fast to slow when weak physical bonds rupture, and, third, terminal nonrecoverable, viscous transport. ϵ/τ vs t may alternatively be plotted as a creep compliance curve (Fig. 2a).

In the Maxwell model, η and **G** are in series and τ varies with time, while ϵ is held constant. In the Voigt–Kelvin model, η and **G** are in parallel, each exercising a damping effect on the other, and an instantly imposed τ is held constant, while ϵ varies with time (Van Wazer *et al.*, 1963). The Maxwell model appears to fit polysaccharides with long linear segments interrupted by junction zones, and the Voigt–Kelvin model appears to fit branched polysaccharides containing junction zones scattered throughout side chains.

ϵ_{max} in a Maxwell fluid is the sum of Hooke's and Newton's laws [Eqs. (4.12) and (4.16); Appendix 3]:

$$\epsilon_{max} = \epsilon + \dot{\gamma}, \tag{4.17}$$

$$\epsilon_{max} = \tau/\mathbf{G} + \tau t/\eta. \tag{4.18}$$

The differential form of Eq. (4.18) is (Kaelble, 1971)

$$d\epsilon_{max}/d\tau = (1/\mathbf{G})(\partial\tau/\partial t) + \tau/\eta. \tag{4.19}$$

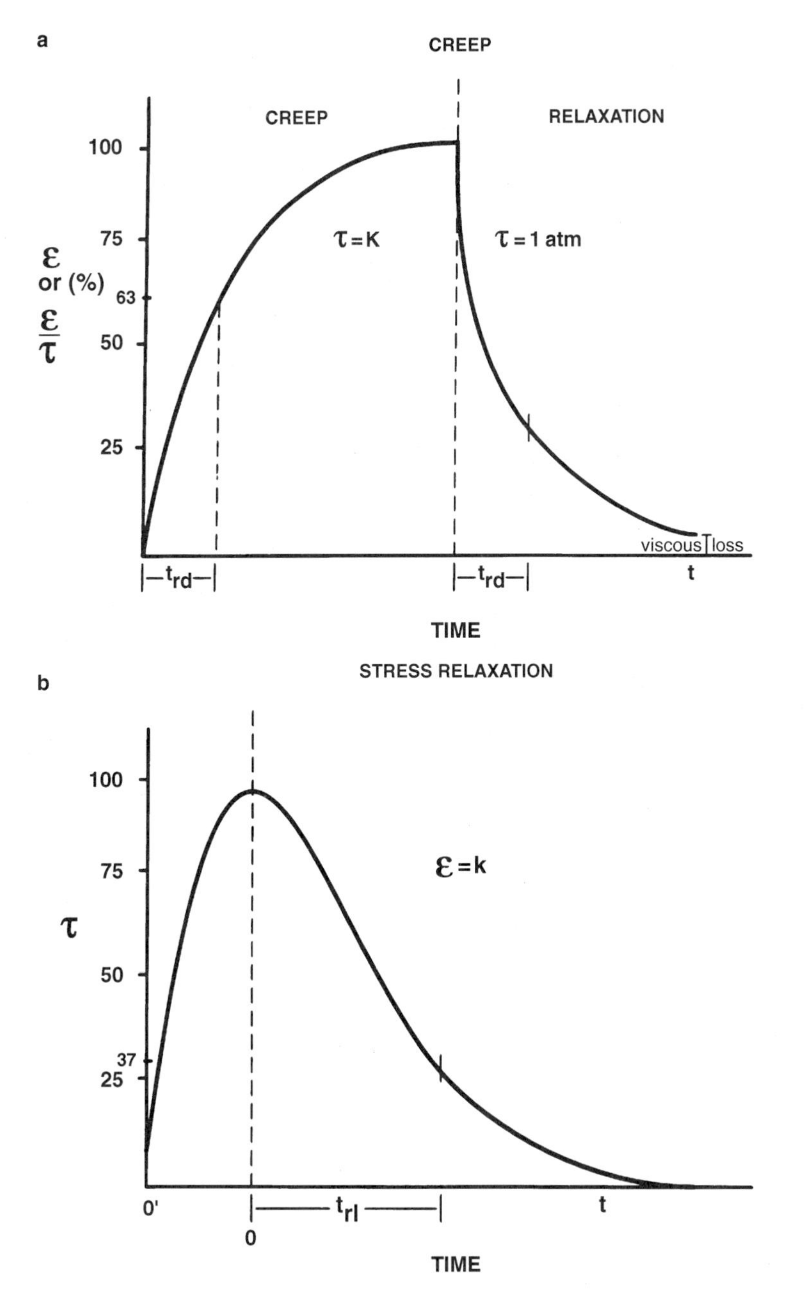

Figure 2 Graphical representation of the Voigt–Kelvin model (a) and the Maxwell model (b) of viscoelasticity. t_{rd} is the retardation time and t_{rl} is the relaxation time.

Under constant τ, $\partial\tau/\partial t = 0$ and Eq. (4.19) is simplified to

$$d\epsilon_{max}/dt = \tau/\eta, \tag{4.20}$$

in harmony with the principle of the consistometer that the "thickest" fluids (highest η_i) suffer the least deformation and travel the shortest distance in any given interval.

The integral form of Eq. (4.18) (Kaelble, 1971) shows that ϵ is an exponential decay function of $t/(\eta_i/\mathbf{G})$. The dimensions of $\eta/\mathbf{G}$ reduce to seconds (Appendix 4) and the equation reaches a limiting $1/e$ ($0.37\epsilon_{max}$) in $t = \eta/\mathbf{G}$ seconds. The retardation time (t_{rd}) is the time required for ϵ_{max} of a Voigt–Kelvin fluid (Fig. 2a) to be reduced to 37% of ϵ_{max} after τ has been removed (Barnes *et al.*, 1989; Seymour and Carraher, 1981). A long retardation time is characteristic of a more elastic than viscous fluid.

In the Kelvin–Voigt test, ϵ_{max} is imposed almost instantly and maintained, while a declining τ is measured from its corresponding maximum (Fig. 2b; Kaelble, 1971):

$$\tau_{max} = \mathbf{G}\epsilon + \eta\epsilon/t. \tag{4.21}$$

From this equation,

$$\epsilon = [(\tau_{max}/\eta) - (\mathbf{G}\epsilon/\eta)]t, \tag{4.22}$$

and by differentiation (Appendix 5),

$$d\epsilon/dt = (1/\eta)(\tau_{max} - \mathbf{G}\epsilon). \tag{4.23}$$

The integral form of Eq. (4.23) shows that τ is a function of $1 - \exp(-t/(\eta/\mathbf{G}))$ and $t = \eta/\mathbf{G}$ seconds is the time required for the fluid to recover 63% $(1 - e)$ of its original shape at $t = 0'$ or, stated differently, to be reduced to 37% of its maximum deformation at $t = 0$ (Fig. 2b). The relaxation time (t^1; t_{rl} in Fig. 2.b) is the time required for τ in a Maxwell fluid to be reduced to 37% of its maximum value at $t = 0$ (Seymour and Carraher, 1981; Barnes *et al.*, 1989). If log ϵ or ϵ/τ vs t is a straight line, the test fluid conforms with the Maxwell model; if the test fluid does not conform, an average t^1 is taken from a spectrum of relaxation times (Mohsenin, 1980).

Deformation $(+\Delta E)$ and recovery $(-\Delta E)$ along dissimilar pathways beget hysteresis. The elastic segment of creep and relaxation can occur at the same rate only when there is no hysteresis. Accordingly, in the absence of hysteresis, t^1 is the time required for a viscoelastic fluid to reach 63% of the maximum deformation under stress.

t^1 is related to $\overline{M}_w$ (Elbirli and Shaw, 1978; Ross-Murphy, 1984) as follows:

$$t^1 = 6(\eta_i - \eta_o)\overline{M}_w / \left[(\text{pi})^2 c_i RT\right]. \qquad (4.24)$$

Silberberg (1989) implied the answer to the question why some polysaccharide gels are flow-reversible and others are not: reversible gels have short t^1 and irreversible gels have infinitely long t^1—meaning the coil–stretch transition of a seemingly irreversible gel has an indefinite interval within which to reverse itself after the deforming stimulus has been withdrawn.

In oscillatory shear rheometry (M. A. Rao, 1992; V. N. M. Rao, 1992), a sinusoidal wave is applied to a dispersion, and the phase amplitudes and differences are measured and related to viscoelasticity. The data yield a complex η (η^*):

$$\eta^* = \mathbf{G}^*/\omega. \qquad (4.25)$$

$\mathbf{G}^*$ is the complex modulus. The shear frequency (ω) oscillates in radians per second between 0 and 90°. There are two components to $\mathbf{G}^*$, viz., $\mathbf{G}'$ (the storage modulus) and $\mathbf{G}''$ (the loss modulus), so that $\eta^* = f(\mathbf{G}') + f(\mathbf{G}'')$. In a dispersion or suspension, $\mathbf{G}'$ is a quantitative measure of elasticity and $\mathbf{G}''$, of viscosity. When $\mathbf{G}' = 0$, the fluid is completely viscous and $+\Delta E$ is dissipated exclusively in viscous transport; when $\mathbf{G}'' = 0$, $+\Delta E$ is stored in elasticity. $\mathbf{G}'$ is a reliable monitor of the sol–gel transition (Fig. 3).

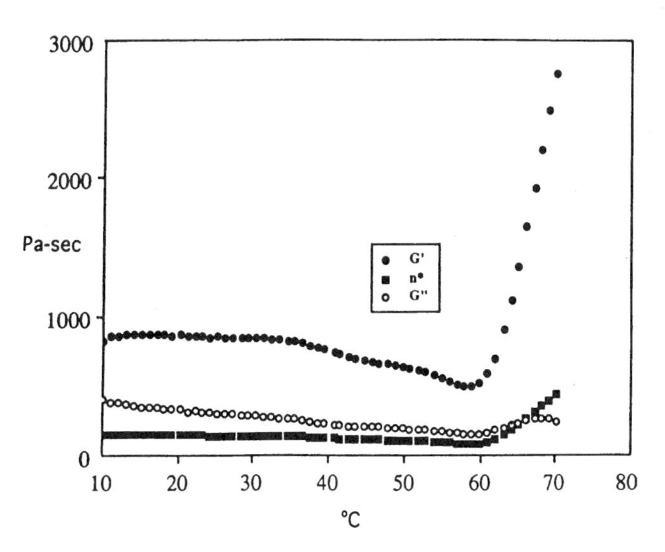

Figure 3 Profile of a complex viscosity (η^*) and the storage ($\mathbf{G}'$) and loss ($\mathbf{G}''$) moduli of a 1% konjac–xanthan–methylcellulose mixture in a 1:1:1 ratio, illustrating the onset of gelation at T_{gel} approximating 60°C.

In the equation

$$\epsilon_{\max} = \epsilon_0 \sin \omega t, \tag{4.26}$$

ϵ_0 is measured as the amplitude of the strain wave at time t_0. Typically, a fixed and a free surface coated with a sample oscillate in or out of phase in proportion to the degree of viscoelasticity. Recalling that $90° = \mathrm{pi}/2$ rad and that the sine of an angle in radians added to pi/2 equals the cosine, the fraction of τ in phase with ϵ may be represented by $\mathbf{G}'(\omega)\sin \omega t$ and the out-of-phase τ, by $\mathbf{G}''(\omega)\cos \omega t$. Then, at any ω, $\tau_{\max}$ is the sum of the fractional τ (Barnes *et al.*, 1989):

$$\tau_{\max} = \epsilon_0[\mathbf{G}' \sin \omega t + \mathbf{G}'' \cos \omega t]. \tag{4.27}$$

J. The Activation Energy of Viscous Flow

Equation (3.35) is a precursor to the Arrhenius equation relating the rate constant [k in Eq. (3.30)] to the probability of molecular collisions (ω) and the activated energy (E_a) of a reaction. Polysaccharide viscous flow is characterized by a modified Arrhenius equation in which η_i/η_o replaces k:

$$\eta_i/\eta_o = \omega \exp(\pm E_a/RT). \tag{4.28}$$

E_a is now the apparent activation energy of viscous flow. ω is called the frequency factor, because it indicates the probability and frequency of the collisions. Severs (1962) made the following observations on dispersions: random polymers change their configuration easily; the most polar polysaccharides require the highest E_a; E_a is less at higher than at lower T; T has a more drastic effect on polar polymers than on nonpolar polymers. With increasing T, random coils expand and a larger solute surface and hence higher η_i (greater resistance to flow) should be effected, but for the heightened Brownian activity that tends to lower η_i. Decreasing T has the opposite effect. Inasmuch as random polymers are easily disfigured by heat, it may be inferred that amorphous polysaccharides decompose in a temperature range lower than do rods, helices, and crystallites. Polysaccharide deploymerization in the presence of oxygen was found to occur at $E_a =$ 50–105 kJ mol^{-1} (11.9–25.1 cal mol^{-1}; Bradley and Mitchell, 1988). The E_a of a cowpea starch gelatinization was reported to be 233.6 kJ mol^{-1} (Okechukwu and Rao, 1996a). E_a was substituted for the time-based empirical criterion used in the jelly trade to classify pectins as slow-set, rapid-set, etc. (Walter and Sherman, 1981).

IV. Size

The size of a polysaccharide normally refers to M or the DP—a restriction that makes no allowance for α, v_{ex}, and v_f—all in a dispersion capable of being many times that of the primary structure at or near θ. Hydration, charge, i, and T are indirect determinants of size.

A. The van't Hoff Equation

With the discovery that polymer molecules in very dilute solution behave similarly to gas molecules, the osmotic pressure generated by the former in solution was equated with the gas pressure [Eq. (3.1)]. The concentration dependence of π in dilute solutions is stated as

$$\pi = RT(c_i)/\overline{M}_n. \tag{4.29}$$

c_i (in grams per liter; c_i/V_i) is the weight of polymer of molecular weight $\overline{M}_n$. This equation governs membrane osmometry of macromolecules, and because its origin is in the combined gas laws, dilution theoretically minimizes errors arising from interparticle interaction. At the same weight concentration, π is high for small molecules, but vanishingly low for polymers, due to the latter's low molarity. The quotient π/c_i is defined as the reduced osmotic pressure, and if π/c_i vs $\underline{c_i}$ (usually a straight line) is extrapolated to $c_i = 0$, the intercept is $RT\overline{M}_n^{-1}$. The equation holds for aqueous dispersions of nonionic polysaccharides and, allowing for Donnan distribution, for ionic polysaccharides dispersed in a dilute concentration of electrolytes. Equation (4.29) is the secondary origin of other equations that explain macromolecular solution behavior.

For rigorous analysis, Eq. (4.29) is expanded to

$$\pi/c_i = RT/\overline{M}_n + \beta c_i + \beta' c_i^2 + \cdots). \tag{4.30}$$

$\beta = 2RT\overline{M}_n^{-1}$ is called the second virial coefficient; it yields the same qualitative information about interaction as $k[\eta]^2$ in Eq. (4.6). Membrane osmometry seldom requires an accuracy to more than $\beta' c_i^2$. Doi and Edwards (1986) define a dilute solution as one in which $\beta = 0$—the ideal condition for accurately measuring $\overline{M}_n$. T_θ is that temperature where $\beta = 0$ (Alberty and Silby, 1992). The fact that β provides information about solute–solute interactions, micellization and demicellization studies are made possible by the use of Eqs. (4.29) and (4.30).

B. Light Scattering

Small, randomly distributed, neutral, noninteracting hydrocolloidal solute undergoes Brownian motion and consequently localized concentration inhomogeneities in a solvent. Oscillating electric fields from a beam of horizontal, plane-polarized, or laser incident light (I_0) transmitted through such an inhomogeneous system induce dipoles in polarizable solute that then scatter light in all directions—uniformly for spherical particles with diameters less than the wavelength (λ_0) of I_0 (Rayleigh scattering), and nonuniformly for particles with diameters approximating λ_0 (Debye scattering). In analogy to Lambert's law, the reduced intensity of the scattered light (I) is stated as

$$I/I_0 = \exp(-\tau xc). \tag{4.31}$$

x is the distance traveled by the scattering beam and τ, called the turbidity, is the scattering equivalent of the absorption coefficient in an absorbing solution. I is of the order of 10^{-6} times the intensity at I_0. Turbidimetry is the difference measurement of I in transmitted light 180° from the source of I_0 (unscattered); nephelometry is the difference measured 90° from the source (scattered light). Quantitative nephelometric data are reliable only to the extent that particle sizes and dispersions are reproducible.

Equation (4.31) provides a quantitative method of analysis and of determining molecular weights and sizes for spherical particles in the 0.01–2-μm-diameter range (Beyer, 1959). Some linear hydrocolloids have small enough dimensions ($< 0.1\lambda$) for them to behave as isotropic scatterers (Sperling, 1986). Large, nonspherical particles scatter light anisotropically, causing a number of waves to arrive at the detector in an identical phase, and other waves to arrive in different phases. I is more intense when waves arrive in an identical phase (constructive inference) and less intense when they arrive out of phase (destructive interference). Constructive interference is most pronounced at forward angles ($\psi < 90°$) where the diffracted rays reinforce each other's intensity: destructive interference, especially acute when particle diameters approximate λ (Debye scattering), is most pronounced at backward angles ($\psi > 90°$) where the out-of-phase frequencies have an attenuating effect on each other. It is easy to understand how polydispersity, through the different size effects, can complicate $\overline{M}_w$ determinations by light scattering. The angular dependence of I introduces the laws of sine and cosine into Eq. (4.31):

$$R_\psi = I_\psi x^2 / \left[V_i I_0 (1 + \cos^2 \psi)\right]. \tag{4.32}$$

R_ψ (in cm^{-1}) is called the Rayleigh ratio and I_ψ is the intensity of the scattered beam measured at ψ. At 90°, $\cos^2 \psi = 0$ and $R_\psi = I_\psi x^2 / V_i I_0$.

A dissymmetry method (Z) has been used to characterize polymers by employing a ratio of I_{45° and I_{135° (with solvent and depolarization corrections):

$$Z = I_{45^\circ}/I_{135^\circ}\ . \tag{4.33}$$

For isotropic scatterers at 45° and 135°, $1 + \cos^2 \psi$ has an identical absolute value that cancels to unity. For anisotropic particles, the concentration dependence of Z complicates the measurements, but this defect is mitigated by plotting a series of Z^{-1} vs c_i points and extrapolating to 0. The intercept, referred to as the reduced specific dissymmetry, is a better index of nonsphericity. Z can be adapted to the determination of $\overline{M}_w$ (Stacey, 1956).

The application of static light scattering to polymers is based on the theoretical equations of Debye (1944, 1947) and the methodology of Zimm (1948). The principles apply equally to polysaccharides (Sorochan *et al.*, 1971). In total intensity light scattering, monochromatic light (436 and 546 nm) at constant T passes through the dispersion and becomes plane polarized; the horizontal beam is scattered in accordance with the equation (Hiemenz, 1986)

$$I_\psi/I_0 = \left\{2\pi^2\left[n^o\left(dn^i/dc_i\right)\right]^2 RTc_i\left(1+\cos^2\psi\right)\right\}/\left\{r^2\lambda^4\left(\partial\pi/\partial c_i\right)\right\}. \tag{4.34}$$

c_i is in grams per centimeter cubed, n^o is the refractive index of the solvent, n^i is the refractive index of the solution or dispersion containing component i, and dn^i/dc_i, the refractive index increment, is the average of the sum of a series of increasing $n^i - n^o$ differences, each difference divided by the respective c_i. It is observed from Eq. (4.34) that the scattering intensity is inversely proportional to the fourth power of λ; this is the reason why scattering at the blue end (436 nm) of the electromagnetic spectrum is far more intense than at the green end (546 nm).

The first derivative of Eq. (4.30) is

$$\partial\pi/\partial c_i = RT\left(\overline{M}_n^{-1} + 2\beta c_i + \cdots\right). \tag{4.35}$$

Inasmuch as molecular weights obtained from light scattering approximate $\overline{M}_w$ more closely than $\overline{M}_n$, the necessary substitutions for $\partial\pi/\partial c_i$ from Eq. (4.35) and $\overline{M}_n$ are made in Eq. (4.34) to give

$$I_\psi/I_0 = \left\{2\pi^2\left[n^o\left(dn^i/dc_i\right)\right]^2 c_i\left(1+\cos^2\psi\right)\right\}/\left\{\lambda^4\mathbf{N}r^2\left(\overline{M}_w^{-1} + 2\beta c_i\right)\right\}. \tag{4.36}$$

$\mathbf{N}/\overline{M}_w$ is the number of scattering particles per gram of solute. Combining all constants into K (mole times centimeter squared per gram squared),

Eq. 4.36 is rewritten

$$R_{\psi} = Kc_i/\left(\overline{M}_w^{-1} + 2\beta c_i\right), \tag{4.37}$$

$$Kc_i/R_{\psi} = \overline{M}_w^{-1} + 2\beta c_i\,. \tag{4.38}$$

Equation (4.38) is a straight line with intercept $\overline{M}_w^{-1}$ and slope 2β (Hiemenz, 1986).

Destructive interference is compensated for by inserting a size- and shape-dependent particle factor [Π(ψ)] in Eq. (4.38):

$$Kc_i/R_{\psi} = \left[1/\Pi(\psi)\right]\left[\overline{M}_w^{-1} + 2\beta c_i\right]. \tag{4.39}$$

For the smallest scattering particles and in the direction of I_0 (transmitted beam) for the largest, $\Pi(\psi) = 1$. For Rayleigh scatterers (Hiemenz, 1986),

$$\tau = \left[32\pi^3 n^{o2}\left(\partial n^i/\partial c_i\right)^2 c_i\right]/\left[3N\lambda^4\left(\overline{M}_w^{-1} + 2\beta c_i\right)\right], \tag{4.40}$$

and formally combining the constants into H, Eq. (4.40) is simplified:

$$Hc_i/\tau = \overline{M}_w^{-1} + 2\beta c_i\,. \tag{4.41}$$

When $\Pi(\psi) = 1$, Eqs. (4.39) and (4.41) are indistinguishable. For the most common conformations (random-coil, rod, and sphere), Π(ψ) is a function of $\sin^{\nu}\psi$ (Stacey, 1956; Allcock and Lampe, 1981; Hiemenz, 1986; Cowie, 1993). Kc_i/R_{ψ} vs $\sin^2(\psi/2) + kc_i$ is a Zimm plot utilizing Eq. (4.39). Zimm plots are applicable to polyanions in their most probable conformation, but with negative slopes (Berth and Lexow, 1991), because of their propensity to ordering effected by charge repulsion (Stacey, 1956). Light-scattering measurements of polyelectrolytes in salt-free solutions and dispersions have borderline accuracy (Dautzenberg *et al.*, 1994).

The adaptation of Z [Eq. (4.33)] to $\overline{M}_w$ determinations presupposes knowledge of conformation. A possibly false assumption of geometry is avoided by deriving $\overline{M}_w$ from a Zimm plot in which extrapolation to $\psi = 0°$ eliminates optical interference and extrapolation to $c_i = 0$ eliminates solute–solute interactions. The two lines converge at the intercept containing $\overline{M}_w^{-1}$, with the slope of the $c_i = 0$ line directly related to R_g^2 and the slope of the $\psi = 0°$ line directly related to 2β. The same quantitative and qualitative information about interaction is deducible from osmometry and light-scattering β (Frank and Mark, 1955).

A light-scattering $\overline{M}_w$ computes to a higher value than an osmometry $\overline{M}_n$, because the statistical origins specify $\overline{M}_w = f(\Sigma s_i w_i^2/\Sigma s_i w_i)$, whereas $\overline{M}_n = f(\Sigma s_i w_i/\Sigma s_i)$, where s_i is the number of particles of i in a weight category and each category is of average molecular weight w_i.

Dynamic light scattering (Everett, 1988; Sun, 1994) is a comprehensive description of certain optical phenomena that are adapted to the measurement by an autocorrelator of short-life fluctuations in I_0 and I_ψ in the λ range 488–635 nm, at different constant angles (0° to ψ), over intervals of 10^{-7} s. The method exploits the high resolution of laser beam frequencies (ω). At a constant temperature, the fluctuations result in a spectrum of ω observed as twinkling by the unaided eye, as the particle scatterers change vibration, rotation, translation, and hence localized densities. The product $I_0 \cdot I_\psi$ is an important parameter that depends on the time scale of the measurements. At any ψ, $I_\psi < I_0$ (the Doppler effect) and the shorter the counting interval $(0, t_1, t_2, t_3, \ldots, t)$, the closer $I(t_1)$, $I(t_2)$, $I(t_3)$, and $I(t)$ are to I_0 at $t = 0$. From 0 to 180°, the corresponding Doppler shift ($\Delta\omega$) is smallest at 0° and largest at ψ. Over the counting interval, I_0/I_ψ vs ω shows a Gaussian distribution curve whose width at half-height ($\Delta\omega_{0.5}$) is related to D through

$$\Delta\omega_{0.5} = D[(4\text{pi}n_o\lambda_o/c_i)(\sin\psi/2)]^2. \quad (4.42)$$

$(4\text{pi}n_o\lambda_o)/(\sin\lambda/2)$ is termed the scattering wave vector (O). $I_0 \cdot I_\psi$, known as the autocorrelation or time correlation function [abbreviated to g(t)], decays exponentially:

$$g(t) = I_0^2 \exp(-t/t_c). \quad (4.43)$$

For monomolecular spherical models,

$$t_c = 1/(DO^2). \quad (4.44)$$

t_c is denoted the correlation time, the relaxation time, or the decay time; $-t/t_c$ is the slope of a computer plot of $\ln g(t)$ vs t. The hydrodynamic radius R_h is related to t_c by thc cquation (Everett, 1988)

$$R_h = (kT/6\text{pi}\eta_o)O^2 t_c. \quad (4.45)$$

Experimentally, t_c is extracted from a series of $\ln g(t)$ at ψ, plotted against O^2 to give a straight line. Deviations arise from size polydispersity, because a different R_h of each fraction yields a different t_c. A graph of t_c vs $1/\psi^2$ yields R_g. For spherical particles, $R_h = R_g$ and for equivalent hydrodynamic spheres, $R_h = 0.665R_g$ (Tanford, 1961).

A statistical analysis of light-scattering data can compensate for polydispersity. In cumulant analysis, $\ln g(t)$ is expanded in a power series and coefficients of the different terms are evaluated against the experimentally obtained t, in search of the closest-fitting average selected by the smallness of the standard deviation. In a histogram method, the experimental t is evaluated against t of a known and variable composition of a small population of discrete particles and the best fit chosen.

On the basis of light-scattering experiments, Burchard (1994) concluded that molecularly dispersed galactomannans could not be prepared, nor was the use of hydrogen-bond breakers satisfactory in accomplishing the dispersion. In his estimation, galactomannan dispersions do not reach a state of thermodynamic equilibrium. By laser diffractometry, Okechukwu and Rao (1996a) found that the ratio of the major and minor axes of ellipsoidal, ungelatinized cowpea starch granules was in the range 1–1.8 μm.

C. The Contour and Persistence Lengths

In the equivalent sphere model, the volume $v_i = 4\text{pi}R_g^3/3$. For spheres, $R_g \approx M^{1/3}$; for random coils, $R_g \approx M^{1/2}$; for rods, $R_g \approx M$ (Ross-Murphy, 1994). Specifically from the Zimm plot (Cowie, 1991),

$$R_g^2 = 3\lambda^2\beta\overline{M}_w/16(\text{pi})^2. \tag{4.46}$$

The ratio of R_g of a linear and a nonlinear polysaccharide with the same composition and DP is a possible index of branching (Zimm and Stockmayer, 1949).

If two lines (l_1 and l_2) are imagined to be the distance of each of the two ends of a primary chain from the center of mass (or from any point on the primary structure), a third line $\bar{r}$ between the two ends forms a triangle with l_1 and l_2: $\bar{r}$ obeys the law of cosines.[15] The random flight model of an unsubstituted linear polymer chain in which there are no steric or energetic hindrances assumes equal bond lengths and all possible bond angles between atoms or segments. This freely jointed chain has a distance $\sqrt{\bar{r}^2}$ (the root mean square end-to-end distance) between its two ends that is theoretically calculable from the law of cosines of the bond angles. All configurations of the primary structure are equally probable at equilibrium, with an average of tightly coiled chains in a Gaussian distribution of $\sqrt{\bar{r}^2}$ (Eisenberg and King, 1977). The effect of a good and a poor solvent on curling and uncurling of a linear molecule (Alfrey *et al.*, 1942; Severs, 1962) is on $\sqrt{\bar{r}^2}$ (Banks and Greenwood, 1975).

The contour length (Flory, 1953) is the fully extended length of a linear polymer, readily visualized in an unsubstituted polyanion reacting to electrostatic repulsion. The real length—the persistence length—is fixed by substitutions, branching, kinking, interactions, etc., that cause the chain to "persist" with a dimension less than the contour length. The contour and

15. Defining $\bar{r}^2$ in terms of the sum of the squares of l_1 and l_2 minus twice the product of $l_1 \cdot l_2$ and the cosine of the angle made by the lines intersecting at the center of mass.

persistence lengths are conceptually equal for a linear, unsubstituted polyanion. The contour length of a polyanion may be more than twice that of the neutral molecule's length with an identical DP (Veis and Eggenberger, 1954).

D. The Mark–Houwink Equation

An equation finding much application to polysaccharides is the Mark–Houwink equation:

$$[\eta] = k\overline{M}^{\nu}. \tag{4.47}$$

k and ν are systemic to each polymer solution and are known to be variable with T, solvent (Alfrey *et al.*, 1942), i, fine structure (Fasihuddin *et al.*, 1988), chain stiffness (Morris *et al.*, 1981), branching (Bahary, 1973), and heterogeneities. There is evidence that ν itself is influenced by $\overline{M}$ (Deckers *et al.*, 1986): Carpenter and Westerman (1975) give this dependence as the reason why Eq. (4.47) is not valid over the entire range of $\overline{M}$.

The determination of $\overline{M}$ with Eq. (4.47) requires prior knowledge of k and ν, that in turn require a homologous series of standards for quantification. Unlike synthetic polymers, the biocolloidal $\overline{M}$ cannot be known with certainty using this equation, because there are no homologous standards with which to predetermine k and ν. For this reason, Eq. (4.47) is referenced against absolute methods (membrane osmometry and light-scattering photometry); however, the equation is an economical alternative to expensive analytical instrumentation. $\overline{M}$ so obtained is an η-average molecular weight ($\overline{M}_v$). Table II lists a number of Mark–Houwink constants reported for some polysaccharides. Constants for other polysaccharides were tabulated by Launay *et al.* (1986) and Lapasin and Pricl (1995).

ν, usually varying between 0.5 and 1.0, may serve as an index of chain conformation. A value of 0.5 is estimated for a random coil in a θ (non-free-draining) solvent (Daniels *et al.*, 1970; Cowie, 1991; Hiemenz, 1986) and 0.8 is estimated in a good solvent (Cowie, 1991). Mitchell's (1979) range is

TABLE II
Mark–Houwink Constants for Some Dispersed Polysaccharides

Polysaccharide	Solvent	k (cm^3/g)	ν	Reference
Amylose	H_2O	1.32×10^{-2}	0.68	Burchard (1963)
Konjac	H_2O	6.37×10^{-4}	0.74	Kishida *et al.* (1978)
Cellulose			0.9–1.19	Stannett (1989)
Guar	H_2O	3.8×10^{-2}	0.723	Morris (1990)

0.5–0.8 for a random coil by the equivalent sphere theory, 1.0–1.2 for a free-draining random coil, and 1.8 for a rod. For a free-draining coil, $\nu = 1$ (Daniels *et al.*, 1970; Hiemenz, 1986). Chain compaction is refleced in $\nu = 0.1$–0.3 (Robinson *et al.*, 1982). K is related to the macromolecular geometry, large for expanded and rigid coils, e.g., cellulose, pectin, and alginate (Lapasin and Pricl, 1995).

E. The Hydrodynamic Volume

By definition, $\overline{M}[\eta]$, canceling to centimeters cubed per mole, is the hydrodynamic volume (Barth, 1986). $\overline{M} \cdot \mathbf{d}^{-1}$ (Glasstone and Lewis, 1960) has identical dimensions:

$$[\eta]\overline{M} = \overline{M}_i \mathbf{d}_i^{-1}. \tag{4.48}$$

$\mathbf{d}_i^{-1}$ (reciprocal density) is defined as the specific volume. $[\eta]\overline{M}$ is adapted to advantage in gel chromatography (synonymous with size exclusion chromatography) whereby size-homogeneous fractions of $\overline{M}$ elute from the same column, each in a unique elution volume (v_{el}), regardless of conformation. Every chromatographic column has a molecular cutoff value or fractionation limit, and molecules whose sizes are above this limit elute in a void volume (v_0) without the equilibrium–disequilibrium–equilibrium times necessary for baseline resolutions: these sizes are "excluded" from the gel micropores. Within the fractionation range, noninteracting molecules theoretically elute in order of decreasing hydrodynamic size; i.e., the larger the $\overline{M}$, the smaller is the v_{el}, because the larger fractions are resolved in the earlier stages of elution. Smaller $\overline{M}$ elute in larger v_{el} and identical sizes elute in an identical v_{el}. A rod elutes sooner than a random coil for macromolecules of equal $\overline{M}$ (Rollings *et al.*, 1983). A gel bed possessing a negative charge hastens the elution of polyanions, because of electrostatic repulsion and also because the polyanions are effectively barred from entering the gel micropores. Polysaccharide gel-bed surfaces have a fractionally negative charge.

If $\log[\eta]\overline{M}$ for each eluted fraction of a standard polysaccharide is plotted against its corresponding v_{el}, a universal graph circumscribed by log $\{[\eta]\overline{M}\}_{standard}$ vs $\{v_{el}\}_{standard}$ may be drawn that fits unknown polysaccharide fractions with identical v_{el}. An unknown $\overline{M}$ may be computed by substitution in the equation

$$\{[\eta]\overline{M}\}_{standard} = \{[\eta]\overline{M}\}_{unknown}. \tag{4.49}$$

The procedure is as follows: a standard graph of $\{[\eta]\overline{M}\}_{standard}$ vs v_{el} is constructed, $[\eta]_{unknown}$ is measured, and $\{[\eta]\overline{M}\}_{standard}$ corresponding to the unknown v_{el} is read from the graph. Inserting the data in Eq. (4.49),

$(\overline{M})_{\text{unknown}}$ is then calculated. The method is accurate for linear chains, but fails for branched and probably stiff polymers (Burchard, 1994).

By another definition, the hydrodynamic volume is the volume of a single particle in solution (Tanford, 1961); or by invoking the unsolvated, spherical molecular model, $(4/3)\text{pi}(\sqrt{\bar{r}^2})^3$ cm^3. The total volume of particles in 1 g of a dispersed solute is $(4/3)\text{pi}(\sqrt{\bar{r}^2})^3$ cm$^3(\mathbf{N}/\overline{M})$. According to Seymour and Carraher (1981), $(\sqrt{\bar{r}^2})^3$ cm^3 is the effective hydrodynamic volume. In terms of the specific volume,

$$\mathbf{d}_i^{-1}\,(\text{cm}^3/\text{g}) = (4/3)\text{pi}\left(\sqrt{\bar{r}^2}\right)^3 \mathbf{N}\overline{M}_i^{-1}(\text{cm}^3/\text{g}). \tag{4.50}$$

$[\eta]$ has the same unit as d^{-1} and it follows that

$$[\eta]_i = \Phi\left(\sqrt{\bar{r}^2}\right)^3 \overline{M}_i^{-1}. \tag{4.51}$$

Φ is the Flory viscosity constant.

For chains (Cowie, 1991),

$$\sqrt{\bar{r}^2} = \sqrt{(6R_g)^2}. \tag{4.52}$$

Short, stiff polymers behave like rods (small r) whose model as a wormlike chain has $\sqrt{\bar{r}^2}$ the same as the contour length and R_g much smaller (Elias, 1979). R_g and $\bar{r}^2$ decrease with curling, compaction, and age.

$\sqrt{\bar{r}^2}$ is related to $\sqrt{\bar{r}_0^2}$ through α (Sun, 1994):

$$\alpha^2 = \bar{r}^2/\bar{r}_0^2. \tag{4.53}$$

Using $\sqrt{\bar{r}_0^2}$ as the unperturbed chain reference and allowing for expansion and contraction,

$$[\eta]_i = \Phi\left(\sqrt{\bar{r}^2}\right)^3 \alpha^3 \overline{M}_i^{-1}. \tag{4.54}$$

At θ, $\alpha = 1$. Making the necessary substitutions for $\sqrt{\bar{r}_0^2}$ and α in Eq. (4.54),

$$[\eta]_\theta = \Phi 6^{3/2} R_\theta^3 \overline{M}_i^{-1}. \tag{4.55}$$

The R_g-$\sqrt{\bar{r}^2}$ relationships permit inferences to be made about macromolecular conformation, i.e., whether or not the molecule is flexible, linear, rodlike, branched, etc. Equations (4.6)–(4.11) differ from Eqs. (4.50)–(4.55) in not having the radius of a sphere as a variable, but having concentration replace size.

$[\eta]_\theta$ of a nonlinear polymer and of its linear counterpart with identical $\overline{M}$ have been rationalized into a practical index of the degree of branching (Bahary, 1973).

The non-free-draining water attached to a polysaccharide molecule and the free-draining water surrounding it travel at different velocities across a boundary whose location is a function of f_c and, therefore, a function of R_g and η_o [Eq. (4.1); Flory, 1953; Tanford, 1961]:

$$f_c/\eta_o = k\left[\mathrm{f}\left(\sqrt{\bar{r}^2}\right)\right], \tag{4.56}$$

$$f_c = k\eta\left[\mathrm{f}(R_g)\right]. \tag{4.57}$$

F. Fractal Dimensionality

Any volume measurement (V) is definable by the cube of a linear dimension:

$$V = r^3. \tag{4.58}$$

Three-dimensional objects occupy space that may similarly be characterized by an equivalency of length times breadth times width. Fractals are such objects: they are irregularly shaped and built upon a constant repeating, microscopic fine structure. A polysaccharide gel, for example, is generated from an almost infinite number of scale-invariant nuclei (Birdi, 1993) multiplied many times into tertiary and quaternary structures. Assuming R_g to be the constant dimension of the fractal nucleus, flocs conform to

$$V = R_g^{\mathbf{D}}. \tag{4.59}$$

D is the fractal dimensionality. If R_g is divided into r sublengths, Eq. (4.59) becomes

$$V = (R_g/r)^{\mathbf{D}}. \tag{4.60}$$

Letting V_t be the total volume of a gel,

$$V_t = (R_g/r)^3 \tag{4.61}$$

and at less than capacity filling,

$$V/V_t = (R_g/r)^{\mathbf{D}}/(R_g/r)^3 \tag{4.62}$$

$$= (R_g/r)^{\mathbf{D}-3}. \tag{4.63}$$

V/V_t of a gel has a meaning resembling that of ϕ_i. The determination of the volume fraction of solute in a sol (ϕ_i) is a simple computation from a series of density-concentration measurements (Walter and Matias, 1989). The incidence of junction zones is proportional to ϕ_i. A random-coil polymer chain in solution has **D** between 1 and 2 (Birdi, 1993).

It is the nature of flocculation and aging to increase the size of the dispersed hydrocolloidal units at the expense of their number. This suggests that it is possible to have a dispersion in which the original, discrete molecules having a low c_i or ϕ_i become larger and highly overlapping—a semidilute solution (Doi and Edwards, 1986). When there is a gelation phase change, ϕ_i approximates 1 as the dispersion approaches the condition of a xerogel. Considering Eq. (4.55), given the magnifying influence of solute–solute interaction on $[\eta]_i$ and since $\overline{M}_i$ is constant, it can be concluded that R_g and R_h in a sol or gel is larger than R_θ and that neither a hydrophobic nor θ environment accomplishes R_g. Walstra *et al.* (1991) discussed some improbabilities of the fractal theory of gelation centering mostly on the oversimplification of the fractal model that culminates in Eq. (4.63).

Surface area and porosity are examples of polysaccharide properties other than gelation that are amenable to fractal analysis.

G. Sedimentation

By accelerating deposition with a slowly rotating ultracentrifuge, Stokes' law [Eq. (4.1)] is modified in uniform circular motion to an equilibrium distribution along the axis of rotation (0 to x), as a function of solute mass (m_i). At higher centrifugal velocities, sedimentation succeeds the equilibrium distribution. Sedimentation equilibrium and sedimentation velocity provide a means to determine M.

The fundamental origin of the relationship between centrifugal force (F), m_i, angular velocity ($\boldsymbol{\Psi}$), and distance from the centrifugal axis (x') is $F = m_i\boldsymbol{\Psi}^2/x'$ (Smith and Cooper, 1957). From this relationship, equations were derived to describe the translational vector during distribution and deposition in a centrifugal field from 0 (the meniscus) to x (the bottom of the cell) along the rotational axis (Williams *et al.*, 1978).

1. Sedimentation Equilibrium

F changes along x according to

$$F = m_i\boldsymbol{\Psi}^2 x. \tag{4.64}$$

If $m_i = M_w$,

$$F = M_i\boldsymbol{\Psi}^2 x. \tag{4.65}$$

Ordinarily a dispersed m_i can be represented as the product of density and partial specific volume ($\mathbf{d}_i v_i$) and v_i, having dimensions of area times length (Ax), any infinitesimal change in Ax ($\partial A\, \partial x$) is accompanied by an infinitesimal change in F (∂F). Substituting in Eq. (4.64) and rearranging,

$$\partial(F/A) = \mathbf{d}_i \boldsymbol{\Psi}^2 x\, \partial x. \tag{4.66}$$

Recalling the definition of pressure ($F/A = \pi$) and substituting for F/A in Eq. (4.66),

$$\partial \pi / \partial x = \mathbf{d}_i \boldsymbol{\Psi}^2 x. \tag{4.67}$$

Simultaneously at every ∂x, there is an infinitesimal change in concentration (∂c_i) and consequently an infinitesimal change in the chemical potential ($\partial \mu$) and the enthalpy change [$\partial(\Delta H_{\text{mix}})$]. From the energy relationships at ∂x, $\partial \mu \approx \partial(\Delta G_{\text{mix}})$. The equation $\Delta G_{\text{mix}} = \Delta H_{\text{mix}} - T\Delta S_{\text{mix}}$ [Eq. (3.18)] is differentiable to

$$\partial(\Delta G_{\text{mix}})/\partial \pi \approx \partial \mu / \partial \pi = v_m. \tag{4.68}$$

Given $v_m = v_i \overline{M}_w$,

$$\partial \mu / \partial \pi = v_i \overline{M}_w. \tag{4.69}$$

At equilibrium, solute transport through the solvent ceases, two gradients affecting $\partial(\Delta H_{\text{mix}})$ and counterbalancing F (viz. $\partial c/\partial x$ and $\partial \pi / \partial x$) are established, and the centripetal and centrifugal forces are equal:

$$M_w \boldsymbol{\Psi}^2 x = (\partial c_i / \partial x)\, \partial \mu / \partial c_i + (\partial \pi / \partial x)\, \partial \mu / \partial \pi. \tag{4.70}$$

One form of Raoult's law [Eq. (2.5)] differentiates to

$$\partial \mu / \partial c_i = RT / c_i. \tag{4.71}$$

An ultracentrifuge is designed to facilitate readings of c_i at different lengths along the rotating axis ($x_0, x_1, x_2, x_3, \ldots, x$); so $\partial c_i / \partial x$ in Eq. (4.70) is experimentally determinable directly. Substitution in Eq. (4.70) of the partial derivatives from Eqs. (4.67), (4.69), and (4.71) and the experimentally determined $\partial c_i / \partial x$ gives

$$M_i \boldsymbol{\Psi}^2 x = (\partial c_i / \partial x) RT / c_i + \mathbf{d}_i \boldsymbol{\Psi}^2 x \cdot v_i \overline{M}_w. \tag{4.72}$$

Rearranging,

$$\overline{M}_w \boldsymbol{\Psi}^2 x = v_i \overline{M}_w\, \mathbf{d}_i \boldsymbol{\Psi}^2 x + (RT/c_i)(\partial c_i / \partial x), \tag{4.73}$$

$$[1/(x\, \partial x)]\, \partial c_i / c_i = \boldsymbol{\Psi}^2 \overline{M}_w (1 - v_i \mathbf{d}_i)/RT, \tag{4.74}$$

$(1/x\,\partial x)$ integrates to $2/x^2$, and $(\partial c_i/c_i)$ integrates to $\ln c_i$. By substitution and transposition,

$$(2/x^2)\ln(c_i/c_o) = \mathbf{\Psi}^2\overline{M}_w(1 - v_i\mathbf{d}_i)/RT, \tag{4.75}$$

$$\ln(c_1/c_0) = \mathbf{\Psi}^2\overline{M}_w(1 - v_i\mathbf{d}_i)(x_1^2 - x_0^2)/2RT. \tag{4.76}$$

c_1 and c_0 are the steady-state concentrations of i at x_1, and x_0 and $(1 - v_i d_i)$ is called the buoyancy factor: if $1 - v_i\mathbf{d}_i$ is positive, there is deposition; if negative, there is flotation (Cowie, 1991). v_i and $\mathbf{d}_i$ are independently measurable. A graph of $\ln(c_i/c_o)$ vs $(x_1^2 - x_0^2)$ has slope $\mathbf{\Psi}^2\overline{M}_w(1 - v_i\mathbf{d}_i)/2RT$ from which $\overline{M}_w$ may be extracted.

2. *Sedimentation Velocity*

In the event of sedimentation, x is the changing solute boundary (the meniscus) distance away from the meniscus at t_0, the initial position of the meniscus. If the weight of one molecule $\overline{M}_w/\mathbf{N}$ is substituted for m_i, Eq. (4.65) is

$$F = (M_w/\mathbf{N})\mathbf{\Psi}^2 x. \tag{4.77}$$

For sedimentation to occur, it is necessary that F exceeds buoyancy only slightly in proportion to $v_i d_i$ (Archimedes principle) and the frictional resistance $(f_c\,\partial x/\partial t)$ [Eq. (4.2)]:

$$(M_w/\mathbf{N})\mathbf{\Psi}^2 x = (M_w/\mathbf{N})\mathbf{\Psi}^2 x\,\mathbf{d}_i v_i + f_c\,\partial x/\partial t, \tag{4.78}$$

$$M_w/\mathbf{N}\mathbf{\Psi}^2 x - M_w/\mathbf{N}\mathbf{\Psi}^2 x\,\mathbf{d}_i v_i = f_c\,\partial x/\partial t, \tag{4.79}$$

$$[1/(\mathbf{\Psi}^2 x)]\,\partial x/\partial t = M_w(1 - \mathbf{d}_i v_i)/\mathbf{N}f_c. \tag{4.80}$$

By definition, the sedimentation constant (S_v) is

$$S_v = [1/(\mathbf{\Psi}^2 x)]\,\partial x/\partial t, \tag{4.81}$$

and a plot of log x vs t is ideally a straight line yielding S_v. The unit of S_v is the Svedberg (10^{-13} s; Hiemenz, 1986). Substituting S_v in Eq. (4.80),

$$S_v = M_w(1 - \mathbf{d}_i v_i)/(Nf_c). \tag{4.82}$$

$f_c = (RT/DN)$ [Eq. (3.27)], and Eq. (4.82) may be rewritten

$$S_v = M_w(1 - \mathbf{d}_i v_i)D/(RT). \tag{4.83}$$

A rearranged Eq. (4.83) takes the form of the Svedberg equation for determining M_w by sedimentation velocity:

$$M_w = RTS_v / [D(1 - v_i \mathbf{d}_i)]. \quad (4.84)$$

Equation (4.84) is absolute, but S_v and D must be known. The concentration dependence of D and S_v necessitates substitution of S_0 from Eq. (4.85) for S_v in Eq. (4.84) (Dautzenberg *et al.*, 1994):

$$1/S_v = 1/S_0(1 + kc_i). \quad (4.85)$$

A plot of the series of experimental S_v^{-1} vs c_i provides slope S_0^{-1}.

Equation (4.86) (Cowie, 1991; Dautzenberg *et al.*, 1994) is an alternative method of determining an unknown M_w that was developed from polymer fractions of known M_w:

$$S_v = k_s M_w^{\nu}. \quad (4.86)$$

D and S_v are related to the van't Hoff equation [Eq. (4.30)]. For polyelectrolytes,

$$(D/S_v)(1 - v_e \mathbf{d}_o) = RT\left(\left(1/\overline{M}_n\right) + \beta c + \beta' c^2 + \cdots\right). \quad (4.87)$$

v_e is the partial specific volume of the polyelectrolyte (Dautzenberg *et al.*, 1994). It is thus possible to procure the second virial coefficient from sedimentation data.

The particle radius (r) of homogeneous suspended solute may be calculated with the use of Eq. (4.88) (Nichols and Bailey, 1949):

$$r = [9\eta_i \ln(x_1/x_0)] / 2\omega^2(\mathbf{d}_i - \mathbf{d}_o)\,\Delta t]^{1/2}. \quad (4.88)$$

Δt is the time elapsing for the solute to travel from x_0 to x_1. Equation (4.88) suggests that the most densely flocculated gel (largest d_i) has the smallest fractal nucleus.

Sedimentation velocity permits the widest possible range of M_w measurements, from 300 to 10^8 Da for dispersions and as high as 10^{14} Da for suspended solute (Dautzenberg *et al.*, 1994).

H. Surface Area

The importance of surface area in colloidal chemistry has spurred many attempts to develop a method of its accurate measurement from physical adsorption processes. All of the methods so far are empirical and attended with difficulty involving surface nonuniformity, polymolecularity, conformational shifts, and multilayer adsorption. Polysaccharide surfaces are seldom solid and uniform, but are more likely to be three-dimensional and contain channels and pores.

Most adsorption data involving physical forces fit the Langmuir equation, stated as (Adamson, 1990; Baianu, 1992a)

$$A_{sp} = n_a a_x . \tag{4.89}$$

A_{sp} is the specific surface area of the adsorbent, a_x is the cross-sectional area of an adsorbate molecule, and n_a is the number of adsorption sites, identical to the total number of molecules adsorbed at $n_a a_x$ without v_{ex} effects, supposing a uniform monolayer thickness of solute at saturation.

At equilibrium, by analogy with an ideal gas, the rates of desorption and adsorption are equal. Before equilibrium, the rate of adsorption of component i is proportional to c_i and the number of unfilled sites; the rate of desorption is proportional only to the number of filled sites. Letting c_i be the solution concentration at any time, c_n be the surface saturation concentration, c_u be the unfilled-sites concentration, $(c_n - c_u)$ be the filled-sites concentration, k_a be the adsorption equilibrium constant, and k_d be the desorption equilibrium constant,

$$k_a c_i c_u = k_d (c_n - c_u), \tag{4.90}$$

$$1/c_u = 1/c_n + k_a/(k_d c_n) c_i . \tag{4.91}$$

c_n, c_u, and c_i are experimentally determinable quantities. Ideally in dilute solution, Eq. (4.91) is linear, giving slope $k_a/(k_d c_n)$. Equation (4.91) is valid in the vicinity of T_c (Kipling, 1965). When all adsorption sites are filled, c_n is constant and the isotherm afterward remains at a plateau concentration.

If a calibrating substance, e.g., a fatty acid, saturates A_{sp}, the total number of adsorption sites is

$$n_a = \mathbf{d} v \mathbf{N}/m. \tag{4.92}$$

$\mathbf{d}$, m, and v are the density, molecular weight, and molecular volume, respectively, of the fatty acid: $\mathbf{d}v$ cancels to grams. c_n is the equivalent of the fatty acid $\mathbf{d}v$. Substituting for c_n in Eq. (4.89), the calibrating equation is

$$A_{sp} = a_0 c_n \mathbf{N}/m. \tag{4.93}$$

The fatty acid cross-sectional area (a_0) has been measured at 20.5 Å^2 (Adamson, 1990).[16] Fatty acids are suitable standards because they order themselves closely packed perpendicularly to a hydrophilic surface.

16. One angstrom equals 10^{-10} m or 10^{-6} cm or 10^{-1} nm.

V. Summary

The size of a topologically linear, random-coil polysaccharide can be determined by adopting the model of an equivalent hydrodynamic sphere, recognizing limitations due to friction, hydration, branching, etc. Dilution mitigates but does not altogether eliminate these effects. Dilute regimes have imprecise boundaries, defined more by practical utility than by rigid physical and numerical criteria.

Polysaccharide molecular weights are determinable by a number of methods that relate directly and indirectly to intrinsic (e.g., primary-chain length) and extrinsic (e.g., electrolyte concentration) factors. Light scattering and osmometry, particularly, can be adapted to quantify absolutely the hydrocolloidal mass and its dimensions. Viscometry is not absolute, but because it is simpler and less rigorous than light scattering and osmometry methods, it has become routine in polysaccharide physical chemistry research. A major shortcoming of the application of viscometry to biocolloids, generally, is the lack of primary standards for determining necessary empirical constants.

A solid polysaccharide surface is measurable by adsorption in a monolayer of a standard compound like a fatty acid. Nonequilibrium accumulation of adsorbate on the polysaccharide solid surface is a function of time.

Ill-defined polysaccharide systems are measurable only as fractal aggregates on whose nuclear surfaces multilayer adsorption of solute can occur.

CHAPTER **5**

Additivity, Complementarity, and Synergism

I. Introduction

Mixed dispersions react by the same number-, weight-, and viscosity-average principles as single-solute dispersions, but combinations may evoke entirely new oral sensations: for example, combinations of starch and carrageenan were evaluated for texture in cream desserts and were found to elicit different responses (Nadison, 1990); Lopes da Silva *et al.* (1993) blended high- and low-methoxyl pectins with locust bean gum in different proportions and also found different responses. Pectin is believed to contribute flavor-release characteristics in yogurts that are superior to those of starch (Hoefler, 1991). Between the extremes of complete phase miscibility and separation, the contribution of each component in a dispersion to the different sensations may be additive, complementary, or synergistic with a variety of compounds. Polysaccharides display mutual antagonism under certain dispersion conditions.

II. Interactions

When two polysaccharides are dispersed in water under gelling conditions, the one more prone to gelation may develop a continuous reticulum throughout the solvent, embedding discrete volumes of the cosolute dispersion. The developed reticulum acquires the properties of the continuous phase polysaccharide and is said to be filled with the cosolute. Alternatively, the solute more prone to gelation may form deposition nuclei for the cosolute (Walter *et al.*, 1978). If the cosolutes are equally capable of gelling, two independent reticula may interpenetrate, giving the gel the appearance of a macroscopically homogeneous dispersion. Random-coil polysaccharides are more likely to interpenetrate than rodlike polysaccharides. What Morris (1992) described as a coupled network may also be possible if segments of each polysaccharide are fused into junction zones.

Dispersed in water, polysaccharides that have the same net charge may separate into two liquid phases (coacervates), each containing water, a major concentration of one solute and a minor concentration of the other, and vice versa. Weak synthetic polyelectrolytes develop gelatinous coacervates and strong synthetic polyelectrolytes develop colloidal precipitates (Eisenberg and King, 1977). Depending on concentrations, the two phases may combine into a solid suspension wherein crystallike bodies formed may act as a stabilizer (Horton and Donald, 1991). These mixed solid dispersions have textural inhomogeneities that are in fact points of weakness, possibly subject to rupture under small stress. Rupture sites in mixed gels affect their mechanical strength and consequently their handling properties, utility, and eating quality (van Fliet *et al.*, 1991).

A. Polysaccharide–Polysaccharide

1. Additivity

Additivity obtains when a measured property, e.g., viscosity, is the sum of the contributions made by each cosolute. Additivity for any pair of compatible polysaccharides fits the following equation developed for cellulose gums (Hercules, Inc., 1980):

$$\log \eta_B = [X \log \eta_1 + (100 - X)\log \eta_2]/100. \tag{5.1}$$

η_B, η_1, and η_2 refer to the viscosity of the blend, the first solute, and the second solute, respectively, and X is the weight percentage of one solute. η_B may also be estimated graphically by reading η_1 and η_2 from the respective vertical axes of Fig. 1 and connecting them with a first line: from the desired η_B on one axis, a second line horizontal to the materials-scale axis is drawn to intersect the first line; from the point of intersection, a third line is drawn parallel to the η axes to intersect the materials-scale axis where the relative percentages of solutes 1 and 2 are read at the desired composition. It is doubtful that blends of more than two polysaccharides have merit, considering the wide spectrum of properties of each.

Blending may also be accomplished by use of the equation (Dow Chemical Co., 1990)

$$\eta_B^{1/8} = X_1\eta_1^{1/8} + X_2\eta_2^{1/8}. \tag{5.2}$$

X_1 and X_2 are the weight fractions of solutes 1 and 2.

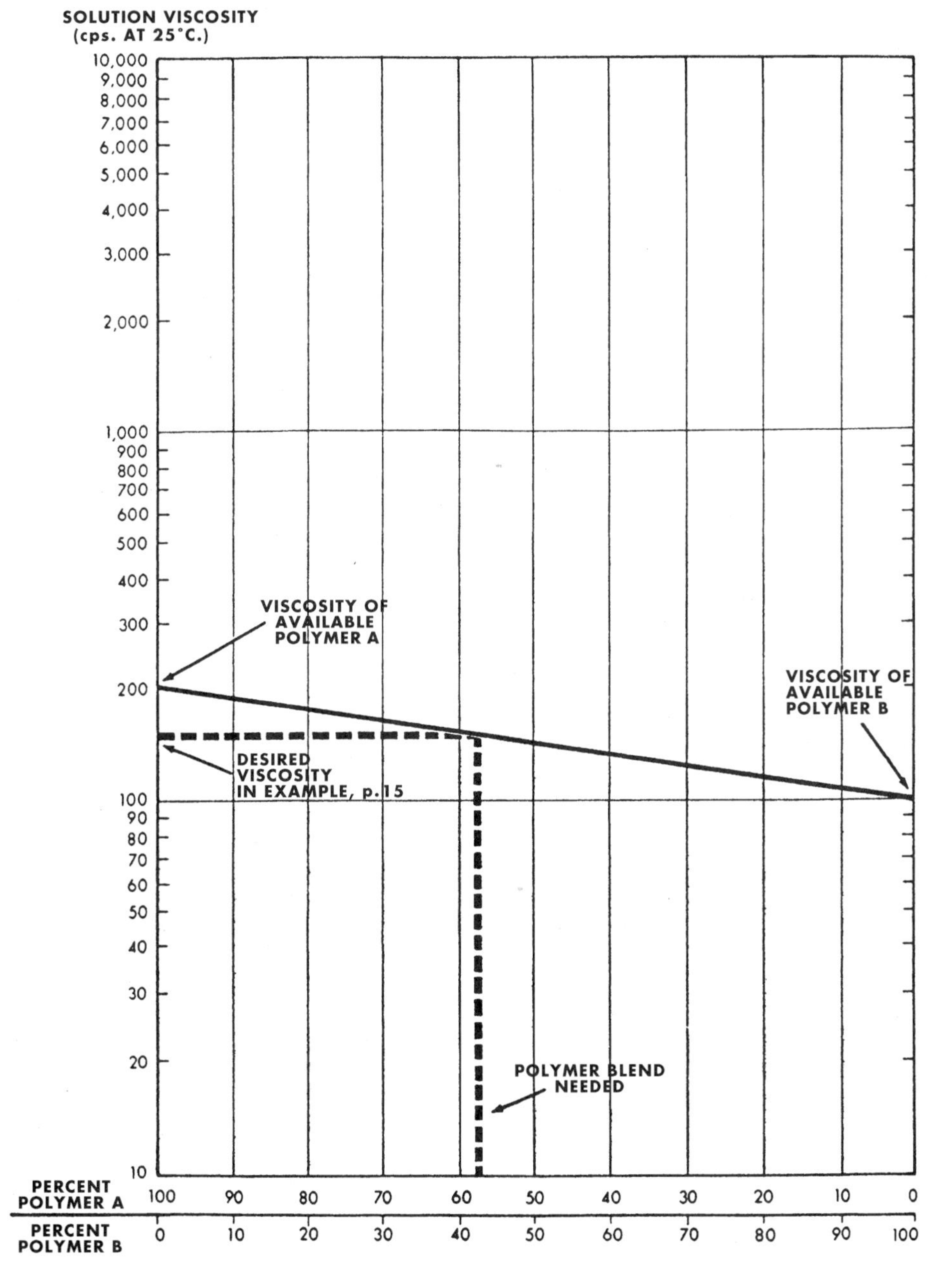

Figure 1 Polysaccharide viscosity blending chart. Reprinted with the permission of Hercules, Inc., Wilmington, DE.

2. Complementarity

One solute may bring to the dispersion an essential property that the cosolute lacks or is deficient in: for example, starch dispersions tend to become undesirably thin by heating—a quality defect in broths and gravies correctible by adding 1–2.9% of thermosetting methylcellulose (Henderson, 1988). The very hygroscopic guar gum is added to starch gels to prevent syneresis after thawing: waxy corn starch was added to curdlan for the same reason (Nakao *et al.*, 1991). Gellan gum shortened the setting time of starch confections from 45 to 24 h (Dziezak, 1991).

3. Synergism

The combined properties of the cosolutes in a single phase may exceed that of the additive property of each solute in separate phases in an identical volume of water: this is synergism. The galactomannans, especially those with the least galactose substituents on the primary structure, display synergism with the broadest array of polysaccharides (Dea, 1989; Sprenger, 1990). The unsubstituted mannan region was implicated in the crosslinking of xanthan and locust bean gums with other gums (Dea and Morris, 1977). Helices and rigid rods increase in viscosity upon the addition of a random-coil polymer (Laurent *et al.*, 1974; Lee and Lee, 1979; Reinhart, 1980).

κ-Carrageenan is synergistic with locust bean gum, but not with guar gum (Moirano, 1977), and with konjac flour containing 85% glucomannan in an optimum range of 30:70–50:50 to a total concentration of 0.6% (FMC, 1989). Some characteristics of the blend are a property of the konjac molecular weight (Kohyama *et al.*, 1993). Carrageenans, pregelatinized starch, and pyrophosphate make a formula for cold-setting milk puddings (Guiseley *et al.*, 1980).

Blends of starch and konjac gum (Tye, 1991) and of some modified celluloses (Hercules, Inc., 1980) are synergistic. A ternary dispersion of 1% methylcellulose and 2.9% starch had almost 2.5 times the viscosity of a blend of 0.5% methylcellulose and 2.9% starch, and approximately 20 times the viscosity of 3.9% starch (Hegenbart, 1989). Methylcellulose–starch viscosity synergism was suggested as a formula to decrease caloric content (Henderson, 1989). Guar gum can increase starch paste viscosity tenfold (Christianson *et al.*, 1981).

Xanthan gum has a remarkable affinity for 1,4-β-D-linked polysaccharides with which it forms thermoreversible (Williams *et al.*, 1991) viscoelastic (Tye, 1991) gels. Deacetylated xanthan is especially synergistic (Tako and Nakamura, 1984; Lopes *et al.*, 1992). Neither xanthan nor locust bean gum by itself is a gelling polysaccharide, but together, there is an approximately hundredfold viscosity increase (Kelco, 1976), ultimately terminating in firm, rubbery gels (Dea and Morris, 1977) at equal concentrations totaling 0.1% (Rocks, 1971). Maximum gel strengths of these blends are in the quotient range of 3/2–2/3 (Kovacs, 1973; Clark, 1988). Xanthan gum does not gel with guar gum, but shows viscosity synergism with it. Xanthan–konjac blends gel synergistically and thermoreversibly (Williams *et al.*, 1991).

Xanthan–locust bean gum gelation is a lock-and-key mechanism that guar gum does not fit (Rocks, 1971). Brownsey *et al.* (1988) came upon evidence of the denaturation (disorder–order transition) of the xanthan helix, and of binding of the stereochemically compatible cellulose chains. Having studied a series of stereochemically compatible synergistic and non-synergistic blends of xanthan and other gums, Morris (1992) concluded that denaturation of the xanthan helix was necessary.

High-methoxyl pectin and alginate, nongelling polysaccharides by themselves, solidify into cold gels without sucrose when the mixture is acidified (Morris and Chilvers, 1984; Toft *et al.*, 1986), although 65% sucrose would otherwise be required to gel the pectin. Low-methoxyl-pectin gels are thermoreversible, but some mixtures with propyleneglycol alginate mixture may be irreversible (Toft, 1982). A study of alginate–pectin mixtures reinforced the belief that regularly spaced atomic groups, arranged in parallel chains, lead to strong, cooperative bonding. Alginate contains blocks of mannurate and α-L-guluronate; pectin is an α-D-galacturonan that is mostly a mirror image of α-L-guluronic acid: the two are synergistic below pH 3.8, upon slowly acidifying the mixture. Presumably, paired segments of galacturonic and guluronic acids containing at least four deesterified monomers each develop strong, cooperative, interchain contacts (Toft *et al.*, 1986).

It had been thought that the molecular basis of synergism was bonding in mixed junction zones, developed as a result of cooperative association (Blanshard and Mitchell, 1979), but current evidence points also to polymer exclusions (Morris, 1990; Doublier and Llamas, 1991) and lock-and-key interactions (Rocks, 1971). Mutual exclusion as well as aggregation of mixed helices (Kelco, 1976) possibly explain the viscosity synergism in dispersion blends of galactomannans and helical polysaccharides. Xanthan with the galactomannans, and xanthan, guar gum, and CMC with starch (Christianson *et al.*, 1981), showed this kind of synergism.

Exact mechanisms may not be completely elucidated, but there is consensus that dissimilarities in localized chemical structures play an important role (Dea, 1989), arguing unavoidable entrapment of one polysaccharide in the gel of the other. Turquois *et al.* (1992) appeared to support the entrapment hypothesis by suggesting that synergistic gelation between κ-carrageenan and locust bean gum results in network coupling.

B. Polysaccharide–Lipid

Lipid matter in fruits and vegetables, omitting the wax coating, is low enough to be considered insignificant. However, many polysaccharide isolates contain traces of lipid matter that are occasionally isolated with them, sometimes in large quantities from oleaginous tissues.

1. The Nature and Properties of Lipids

Naturally occurring food lipids are composed of short- and long-chain fatty alcohols, acids, esters, and conjugates: glycerol is the dominant alcohol. Almost invariably, they consist of straight-chain fatty acids, which help to predict their physical state on the basis of incremental additions of mass 14 ($-CH_2-$). Shorter chains are more likely to be liquid (oil) and longer chains are more likely to be solid (fat). Physical properties also depend on the degree of fatty acid unsaturation and on the relative fatty acid positions on glycerol. Animal lipids tend mostly to be saturated fats and plant lipids tend to be unsaturated oils. Fatty acids containing up to 10 carbons (capric acid) are water-soluble, emulsifiable, and steam-distillable.

A single lipid molecule is small by colloidal standards, but micellization augments the unit particle size to colloidal dimensions. All lipid compound molecules are hydrolyzable in water emulsion with varying degrees of ease. Hydrolysis is accelerated by strong acid and alkali. The enzyme lipase hydrolyzes fatty acid esters to glycerol and fatty acids; lipoxydase oxidizes the unsaturated sites to hydroperoxides. Some fatty acids are hydroxylated, and across the hydrophobic spectrum, lipid polarizability ranges from hydroxylated fatty acids and polyunsaturated monoesters to fully saturated triglycerides.

2. Conjugates

Lipids complex amylose to form low-melting, reversible conjugates that prevent starch hydration, inhibit granule swelling, minimize leaching, delay or eliminate retrogradation, deprive it of its ability to gel, reduce digestibility, and make the molecule resistant to α-amylase (Howling, 1980; Galloway *et al.*, 1989; Biliaderis, 1992). They have a profound effect on wheat-starch gel strength, syneresis, pasting peak, and consistency (Takahashi and Seib, 1988). Starch–lipid complexes are believed to influence the baking characteristics of cereals by elevating T_{gel} (Belitz and Grosch, 1987).

Carrageenan and phospholipids stabilize milk fat by a sulfate–amine reaction (Yalpani, 1988). Allegedly, the "rich" sensation of a 3.5% butterfat milk can be achieved in 1% low-fat milk by adding ι- and κ-carrageenan at 0.02–0.04% concentration. The texture and appearance of fluid skim milk ($<0.5\%$ butterfat) can be improved similarly (Moirano, 1977).

Glycerol stabilized a dispersed protein, while fatty acids accelerated its denaturation (Buttkus, 1970). Emulsan, a naturally occurring bacterial lipoaminopolysaccharide containing approximately 15% lipid and a protein fraction, is strongly surfactant.

C. Polysaccharide–Metal

Many naturally occurring ionic polysaccharides are mixed salts of alkali, alkali–earth, and transition metals with different insolubilities. Salts of alkali metals are invariably soluble. Sodium, the most ubiquitous alkali, possesses a single valence electron, large atomic and ionic radii, and very low ionization potential. Na^+ hydrates in aqueous solution and retains its coordination water in the solid state. Prior to use, native polysaccharide salts are usually converted to the sodium form whence they acquire functionality.

Alkali–earth metals (calcium, barium, and magnesium) complex with polysaccharides extensively (Reisenhofer *et al.*, 1984). Calcium has a smaller atomic and ionic radius than does sodium and, because it has two valence electrons, it is endowed with greater polarizing and bonding ability than Na^+. Ca and Ca^{2+} easily form insoluble complexes with oxygenated compounds. Polysaccharide salts of alkali–earth metals are generally insoluble.

Aluminum is an amphoteric element that acts as a nonmetal in alkali and develops a hydrated gelatinous aluminate of a species $[Al(OH)_4]_n^-$. As a result of this reaction, certain suspended matter including polysaccharide polyanions coprecipitates by entrainment. This element, applied in atomic or ionic form, is a common technique for commercial isolation of pectin. In acid, Al^{3+} supposedly neutralizes polyanions to yield the aluminum salt. After precipitation of the pectin–aluminum complex, the metal ligand is removed by acidification and washing.

Boron is devoid of metallic character: in water, it generates weakly acidic boric acid $[B(OH)_3]$. This hydroxide bonds covalently with vicinal (neighboring) hydroxyl groups to form negatively charged, acidic complexes.

Copper is an element of the transition series, much less reactive than the alkali and alkali–earth metals. It possesses one principal valence electron, but is capable of existing in more than one valence state (+1, +2, +3) as a result of electron transfer from inner orbitals. The multiplicity of valence states permits Cu to enter into chemical complexation reactions with inorganic and organic compounds alike. Many Cu solutions are colored blue. Fehling's reagent, used to test for reducing sugars, is a water solution of $CuSO_4$, sodium hydroxide, and a sequestrant (potassium tartrate). The reagent loses color in a positive reaction, e.g., in a test for aldehyde and ketone sugars in which Cu^{2+} is reduced to Cu^+, as the sugar is oxidized.

Iron is a nonamphoteric, transition element with the ability to exist in two oxidation states—Fe^{2+} (ferrous) and Fe^{3+} (ferric). A positive reaction to alkaline ferric chloride is an indication of the presence of hydroxyl groups with which Fe^{2+} forms colored complexes. Stable copper and iron chelates are common in plants. *In vitro*, Cu and Fe are components of solvent systems for cellulose (Jayme and Lang, 1963).

D. Cyclodextrin and Amylose Clathrates

Hydrogen atoms are located on the inside of the cyclodextrin ring, creating hollow hydrophobic interiors capable of electrostatically binding linear, similarly hydrophobic compounds. The interaction is referred to as molecular encapsulation: the cyclodextrin is the "host" and the complexed molecule is the "guest" compound. This host–guest reaction can advance by polymerization to supramolecular structures (Harada *et al.*, 1993). The complexes decompose by heating to 240–265°C. Cyclodextrin–lipid complexes are surfactants (Shimada *et al.*, 1992). The amylose helix also bears the pyranose hydrogen atoms on the inside and the hydroxyl groups on the outside of a spiral primary chain, which makes the interior surface strongly hydrophobic and the outside strongly hydrophilic (Freudenberg *et al.*, 1939; Kerr, 1950). As a result, the hollow interior of the amylose helix is conducive to fatty acid complexation. The net hydrophobicity of most of these complexes confers a degree of insolubility and crystallinity on them.

E. Polysaccharide–Protein

Like polysaccharides, proteins impart texture and structure to foods. Protein–polysaccharide blends may assume physical, ionic, or covalent character, may be soluble or insoluble, and may sometimes exhibit synergism. The characteristics of the blend are as much a response to the reaction environment as to the properties of each ligand.

1. *The Nature and Properties of Proteins*

Protein molecules fall within the limits of colloidal dimensions and are therefore subject to the same interfacial forces as polysaccharides. They too are amphilphilic—existing as zwitterions (internal salts) in water at neutral pH and in the solid state, and reacting chemically and physically with ionic and some nonionic molecules, often specifically (Hart *et al.*, 1992). Their complexes have different bond strengths along with possibly different conformations (Dickinson and Euston, 1991a, b; Mackie *et al.*, 1991).

The amino acids comprising proteins are held together by the peptide bond (–OCNH–) in open (linear), cyclic (spherical), or branched configuration. Proteins and polypeptides, are more compact and dimensionally stable than polysaccharides, because of the inherent preponderance of ionic and disulfide bridges. The peptide bond is more resistant to chemical hydrolysis than the glycoside bond, but is nevertheless vulnerable to prolonged heating in acidic and basic media and to enzymes. Advanced hydrolysis yields a mixture of short-chain peptides with various DP. In all its reactions with polysaccharides, the protein may be negatively or positively charged or may have a net charge of zero (at the isoelectric point; pI). Many more neutral and charged polysaccharides do not engage in any visible reaction with proteins (Dickinson and Euston, 1991a). Heating denatures proteins and, like polysaccharides, they decompose before reaching T_m. Proteins are protective colloids and emulsifiers more efficacious than polysaccharides. The sulfated polysaccharides are the most chemically reactive with proteins.

2. *Polysaccharide–Protein Blends*

According to Stainsby (1980), there are three kinds of interactions between a polysaccharide and a protein, viz., electrostatic, specific ion, and covalent. The first mechanism is characterized by pH and electrolyte sensitivities; the second mechanism is also pH- and electrolyte-sensitive and additionally is compositional; the third mechanism is heat-irreversible. The order of acidification can determine whether or not the resulting complex remains dispersed or precipitates. Proteins equilibrate as polycations in acidic media (below pI) where they can be stablized by polysaccharide polyanions (Ganz, 1974). Mixing at an initially high pH fixes the protein with a net negative charge, leading to mutual repulsion and deposition when a polyanion is added at that pH. Too low an initial pH depresses polyionization and flocculates casein; consequently, there may be no reaction. Sulfated polysaccharides form soluble complexes with proteins above and below the pI.

Pectate, alginate, and CMC have held proteins dispersed under conditions that might otherwise have caused precipitation (Imeson *et al.*, 1977). Polysaccharide stabilizers, in the order of decreasing thermodynamic compatibility with proteins, are pectin > CMC > alginate > gum arabic > dextran (Tolstoguzov, 1986).

The emulsion stability of gum arabic (at constant nitrogen content) was found to increase with the polysaccharide molecular weight (Dickinson *et al.*, 1991a, b). The emulsifying power is attributed to preferential adsorption of the protein moiety that bonds hydrophobically with oil, leaving the hydrophilic polysaccharide moiety free to protrude from the droplet surface into the surrounding aqueous medium (Randall *et al.*, 1988).

Through Ca^{2+} mediation, polysaccharide–protein reactions can occur above the pI where the two ligands are polyanions (Guiseley *et al.*, 1980; Dalgleish and Morris, 1988; Hart *et al.*, 1992). These anionic complexes are differentiated from complexes of paired opposite charges by their indifference to pH.

The covalent bonding of polysaccharide and protein may be induced by heating a mixture under conditions of low a_w, below the denaturation temperature (Dickinson, 1993); these are the most heat and freeze–thaw stable protein–polysaccharide complexes (Stainsby, 1980).

Although critical concentrations of a polysaccharide may enhance the emulsifying and stabilizing ability of a protein, small additions may have the opposite effect (Cao *et al.*, 1990). The discovery of destabilization by 0.05% xanthan on caseinate led Dickinson and Euston (1991a) to remark that small changes in polymer structure or solvent conditions can easily tip the balance in favor of depletion or bridging flocculation.

Polysaccharide–protein combinations occasionally display synergism, as, for example, that between gelatin and gellan (Shim, 1985); this has enormous implications for fabricated fibrous food texture. Some popular combinations are gelatin–agar, casein–alginate, and casein–pectin. Gelatin–agar dispersions are flocculated at pH 3 for use in sherbets and ices (Meer, 1980a); casein–alginate systems make room-temperature dessert gels (Cottrell and Kovacs, 1980). The most important protein–polysaccharide reaction in industrial food processing is that of milk and κ-carrageenan (sodium salt); the latter is added prior to evaporation to obstruct casein aggregation, as calcium (Lin, 1977) and casein concentrations increase with water loss. The ordinarily nongelling λ-carrageenan is indifferent to calcium, but develops a gel with casein in cold milk (Guiseley *et al.*, 1980).

Cryostabilization (Levine and Slade, 1988) is conceptually a mechanism of insuring stability of a dispersed protein by inclusion of small molecules that raise T_g above ordinary storage temperatures where the protein normally resides in a denaturing environment. A starch hydrolysate offered such protection to frozen, comminuted protein dispersions through this mechanism (Buttkus, 1970). The old thinking on cryoprotection was that water flowed osmotically into the protein micelles, not only creating a microenvironment unfavorable for denaturation, but performing a eutectic role also, with the result that the protein dispersion remained in the sol state, once the storage temperature did not fall below the eutectic temperature.

3. Phase Diagram of a Polysaccharide–Protein Blend

Dynamic systems are characterized by rates of change—the definition of d and ∂. dP is positive if a dependent variable increases, whether or not an independent variable increases, and negative if the dependent variable decreases, whether or not an independent variable decreases. At any maximum or minimum, d is zero and dP is constant. Information about the velocity of the change (the rate of change of d) is extracted from the second derivative of the equation that governs the relationship between the independent and dependent variables: this differentiation of d (yielding the second derivative denoted d^2) provides information about maxima and minima. Phase diagrams are used to portray differential changes in ternary dispersions of water, polysaccharide, and protein.

If T of an aqueous polysaccharide (1)–protein (2) dispersion is slowly raised, an upper T_c is eventually reached, where the condition is imposed that $\Delta G_{1,2} < (\Delta G_1 + \Delta G_2)$ and the cosolutes are miscible in all proportions. Below T_c, the blend is either in separate phases or in stable, metastable, or unstable equilibrium of partially miscible phases, depending on the concentration range. If there is phase separation, one phase contains a major concentration of the polysaccharide and a minor amount of protein, and the other phase vice versa. Two such phases in equilibrium are called conjugate solutions (Glasstone and Lewis, 1960). In the mixed dispersion, $\phi_1 + \phi_2 = 1$, and a graph of ΔG_{mix} vs ϕ_i at each T below T_c (Fig. 2) shows two stable concentrations at ϕ_1' and ϕ_1'', where ΔG_{mix} is minimum, and one unstable concentration at ϕ_1, where ΔG_{mix} is maximum. Metastability prevails in the

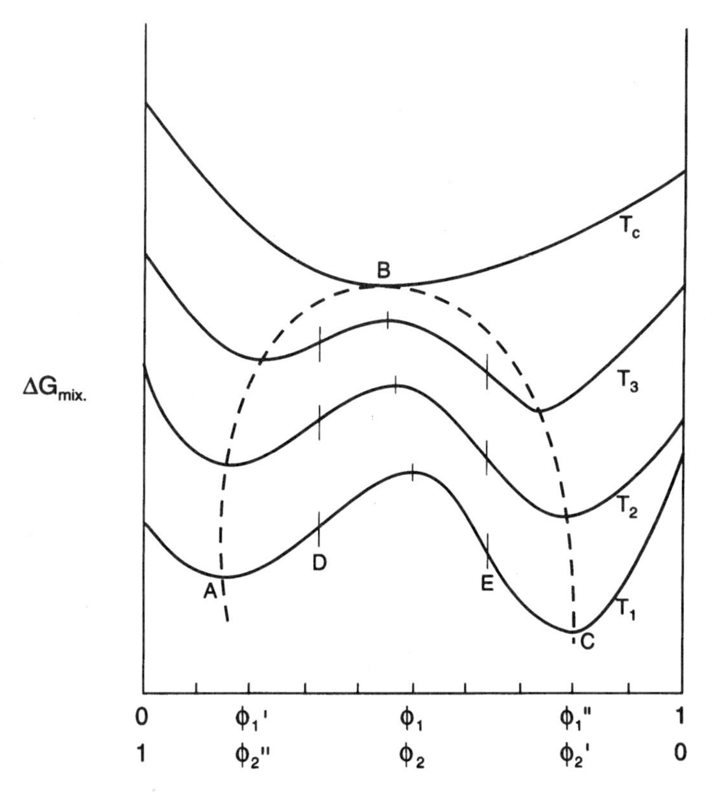

Figure 2 Typical phase diagram of an aqueous polysaccharide (1)–protein (2) dispersion showing the Gibbs free energy as a function of the volume fraction (ϕ) of each, at different temperatures from T_1, where the dispersion is metastable, to the critical solution temperature (T_c), where the two components are miscible in all proportions. *ABC* is the spinodal curve: *DBE* (not connected) is the binodal curve.

intervening range on either side of ϕ_1 between ϕ_1' and ϕ_1''. The asymmetry of the lines reflects the size and shape disparities between components 1 and 2. At the A and C minima and B maximum, $d(\Delta G_{\text{mix}})/d\phi_1 = 0$; from $\phi_1 = 0$ to A on the ϕ axis, $d(\Delta G_{\text{mix}})/d\phi_1 < 0$ (negative), until A; thereafter, $d(\Delta G_{\text{mix}})/d\phi_1 > 0$ (positive) to the maximum at D. As a result of the negative-to-positive change from 0 to D, $d^2(\Delta G_{\text{mix}})/d\phi_1^2 > 0$ (positive). From D to B, $d(\Delta G_{\text{mix}})/d\phi_1$ remains positive, d declines to 0 at B, and declines further to E (negative slope); from D to E, $d^2(\Delta G_{\text{mix}})/d\phi_1^2 < 0$. From E to C, $d(\Delta G_{\text{mix}})/d\phi_1$ remains negative, but the rate change is positive to $\phi_1 = 1$, resulting in $d^2\, \Delta G_{\text{mix}}/d\phi_1^2 > 0$ from C to $\phi_i = 1$. D and E are inflection points, where $d^2(\Delta G_{\text{mix}})/d\phi_1^2 = 0$, inasmuch as the rate change on one side of each is opposite in sign to the rate change on the other side. For ϕ_2 in the opposite direction, the signs are reversed. The conjugate solutions are completely miscible at any concentration at and above T_c.

The events typifying Fig. 2 may be stated nonmathematically as follows: an aqueous dispersion of a polysaccharide and a protein not exceeding a total concentration of 4% (Tolstoguzov, 1986) is relatively most stable at the compositions ϕ_1' and ϕ_1'' and least stable at ϕ_1; metastability can be expected in the vicinity of the concentrations at D and E.

The locus connecting the minima ABC in Fig. 2 is the cloud point or binodal curve and DBE (not connected) is the spinodal curve (Cowie, 1991).

The events outlined in Fig. 2 are expressed in Eq. (5.3), derived from a series of substitutions and differentiations of ΔG_{mix} with respect to the solvent's partial molal content (Cowie, 1973, 1991):

$$d(\Delta G_{\text{mix}})/d(\mathbf{N}n_o) = RT_c\left[\ln(1-\phi_2) + (1 - 1/x_n)\phi_2 + \chi\phi_2^2\right], \tag{5.3}$$

$$\chi = \Delta H_{\text{mix}}/RT_c\mathbf{N}n_o\phi_i. \tag{5.4}$$

Nn_o is the number of solvent molecules, x_n is a dimensionless DP function, and χ is the Flory–Huggins interaction parameter calculable from osmometry data (Ulrich, 1975):

$$\pi/RTc_i = (1/M_i) + \left[\mathbf{d}_o/\left(M_o\mathbf{d}_i^2\right)\right](0.5 - \chi)c_i + \cdots. \tag{5.5}$$

M_i and M_o are the solute and solvent molecular weights, respectively.

The second virial coefficients in Eqs. (4.30) and (5.5) are related via the equation

$$\beta = \left[\mathbf{d}_o/\left(M_o\mathbf{d}_i^2\right)\right](0.5 - \chi). \tag{5.6}$$

When there is no interaction, i.e., when $\beta = 0$, $\chi = 0.5$, the maximum theoretical value. The experience with synthetic polymers is that χ approximates 0.5 in poor solvents and is lower in good solvents (Cowie, 1973). Invalid assumptions of the theory spawning Eq. (5.3) are discussed by Cowie (1991).

III. Antagonism

In antagonism, the combined property imputed to cosolutes in a dispersion is less than the sum of the properties imputed to each solute. If for any reason, e.g., geometry precluding parallelism and identical charge instigating repulsion, the cosolute segments are more cohesively than adhesively attracted, each cosolute may gel independently in a separate saturated phase containing a minor proportion of the other. The two phases may remain a liquid separated by a sharp interface under appropriate conditions of concentration, pH, i, etc. Coacervation is an antagonistic event, according to Dickinson (1993); it differs from thermodynamic instability in that it produces one phase that is almost entirely solvent.

Although mutual repulsion can rationally explain antagonism (Robb, 1986), it cannot do so between antagonistic gelatin (positively charged) and amylopectin (zero charge) (Grinberg and Tolstoguzov, 1972), CMC (negatively charged) and nonionic gums (Hercules, Inc., 1980), neutral dextrin and neutral amylose (Kalichevsky *et al.*, 1986), and possibly neutral locust bean gum and low-methoxyl pectin (negatively charged) (Lopes da Silva *et al.*, 1993). Moreover, amylose and amylopectin exist commensally at higher concentrations in starch granules in various ratios, but when isolated, they are mutually incompatible at moderate concentrations of each totaling 2.5 w/w% (Kalichevsky and Ring, 1987) to 4% (Morris, 1990). The amylose:amylopectin range of incompatibility is 15:85–22:78. Below approximately 15% amylose, amylopectin dominates the continuous phase; inversion occurs above approximately 15% (Doublier and Llamas, 1993).

The introduction of Na^+ ($> 10^{-3}$ M) or Ca^{2+} (5×10^{-4} M) had an antagonistic effect on the xanthan–guar gum interaction. This antagonism was thought to be due to the cationic screening of the xanthan charge and the resulting conformational change (Clark, 1988).

IV. Summary

Polysaccharides interact additively, complementarily, synergistically, and antagonistically with other food molecules. Heated polysaccharide–protein blends tend to be most stable at a low volume fraction of either cosolute; they become increasingly unstable as the concentrations approach equality. The miscibility of phases depends strongly on temperature. Binary and ternary dispersions alike manifest incompatibility by metastability, phase separation, and deposition. Stability can be prolonged by adjusting the volume ratios and storage and processing temperatures within critical limits.

CHAPTER **6**

Thermal Processing

I. Introduction

The time–temperature integral is the singularly most effective stimulus on the polysaccharide disperse system, from the mildest process that insures safety and elementary dissolution to the severest process that initiates chemical decomposition. On the lower response scale, gelatinization and swelling are primary occurrences; on the upper response scale, chemical dehydration, pyrolysis, and resynthesis generate higher $\overline{M}$ species in the involatile phase (Fagerson, 1969) and flavors, aromas (Teranishi *et al.*, 1992), colorants (Vercellotti and Crippen, 1991), and a host of other small organic molecules (Horton, 1965) terminally. Fresh bread, pastries, roasted peanuts, etc., owe their sensory appeal to pyrolysis.

The thermally induced changes may not always be beneficial. Polysaccharide decomposition products are generally considered safe for human consumption, but traces of potentially deleterious compounds, e.g., phenol (Byrne *et al.*, 1966) and acrolein (Walter and Fagerson, 1968), may appear. The γ-lactone of 4-hydroxy-2-pentenoic acid—a vapor-phase constituent of the pyrolysate—has been implicated in antibiosis (Oxford, 1945).

II. Atmospheric and Retort Processing

With few exceptions, polysaccharides are processed for food in a moderately high a_w environment from 100°C and 1 atm to 121.1°C and >1 atm (retorting, ultrahigh temperature pasteurization, and extrusion). Mindful of the combined gas laws ($\pi V_i = n_i RT$), the polysaccharide (n_i) undergoing thermal processing in enclosed space (V_i) must be subjected to high stress (π). Heating to high temperatures and pressures in a closed, evacuated container causes desolvation and demixing of the compacted macromolecules, as water volatilizes into the headspace. Desolvation and demixing of the food substrate are a condition of $-\Delta S_{mix}$ (Elias, 1979). Cooling reverses the process.

The thermophysical events in a can in the retort (dissolution, hydration, dehydration, gelatinization decrystallization, defibrillation, curling, uncurling, etc.) obviously must be complex. Charge superimposes electrostatic and electrokinetic reactions on the thermophysical processes. Broken-curve profiles for some polysaccharide foodstuffs manifest a transition from conduction to convection heating, as a tenuous, reversible suprastructure reverts to a liquefied mass under the influence of $+\Delta H_{mix}$.

The effect of heat on the polysaccharide–water interaction in several dispersions and suspensions was studied by comparative viscometry and rheometry (Tables I–IV). The polysaccharides were the purest manufacturers' grade laboratory washed and dried before dispersion. The dispersion concentrations were below c^* to accommodate capillary viscometry, and the suspension concentrations were above c^* to accommodate rheometry. It is seen in Tables I and II that the cellulose derivatives made the most stable dispersions and the propylene glycol alginate made the least. Dispersions of the neutral polysaccharides were more stable than those of the ionic polysaccharides. From Tables III and IV, it can be argued that suspensions benefit

TABLE I
Ratio of Intrinsic Viscosity of Polysaccharide Dispersions Heated ($[\eta]_h$) for 15 min at 121°C and the Corresponding Unheated ($[\eta]_c$) Controls[a]

Polysaccharide	$[\eta]_h/[\eta]_c$
Methylcellulose	1.0
Na-CMC	1.0
Hydroxypropylmethylcellulose	0.9
Hydroxyethylcellulose	0.9
Konjac gum	0.9
Locust bean gum	0.8
Guar gum	0.8
Xanthan gum	0.7
Gellan gum	0.7
Carrageenan	0.4
Baking pectin	0.3
Algin	0.3
Low-methoxyl pectin	0.2
Propyleneglycolalginate	0.0

[a]The highest dispersion stability is indicated by unity.

TABLE II
Ratio of the Slope of η_i vs c_i for Polysaccharide Dispersions Heated (β_h) for 15 min at 121°C and the Corresponding Unheated (β_c) Controls

Polysaccharide	β_h/β_c
Methylcellulose	0.8
Na-CMC	1.0
Hydroxypropylmethylcellulose	1.1
Hydroxyethylcellulose	0.5
Konjac gum	0.6
Locust bean gum	0.6
Guar gum	0.5
Xanthan gum	0.6
Gellan gum	0.7
Carrageenan	0.2
Baking pectin	0.1
Algin	0.3
Low-methoxyl pectin	0.0
Propyleneglycolalginate	0.1

from heterogenous solid-phase stabilization. The approximately 11% CMC coat on Avicel undoubtedly added thickening power as much as did hydration of the heat-disordered microfibrils.

Rao *et al.* (1981) reported the primary structure of guar gum and CMC to be quite stable after heating at 210–260°F for 5–20 min, but there was a loss of η_a. The results of this heating study suggested possible coil–stretch deformation of an equilibrium structure in the gums beyond their elastic limit and an infinitely long t^1 after cooling. Coil–stretch deformations can

TABLE III
Ratio of Apparent Intrinsic Viscosity of Heated ($[\eta_a]_h$) to Control ($[\eta_a]_c$) for Some Polysaccharide Suspensions

Polysaccharide	$[\eta_a]_h/[\eta_a]_c$
Avicel FD-100	2.5
Avicel RC-591	1.5
Amylopectin	1.7
Corn starch	3.0
Amylose	27.4

TABLE IV
Ratio of Slope (β) of Heated (*h*) to Control (*c*) for Some Polysaccharide Suspensions

Polysaccharide	β_h/β_c
Avicel FD-100	0.9
Avicel RC-591	1.1
Amylopectin	0.9
Corn starch	7.7
Amylose	1.3

also explain viscosity loss in flow-through (tubular and plate) pasteurization and birefringence.

A distinction is made between dispersion stability and chemical bond stability: the former refers to the tenacity of a reversible tertiary structure in its dispersion medium (solid, liquid, or gas) wherein any conformational shift is theoretically transient and finite, albeit in a possibly long interval. Chemical instability involves decomposition of the primary structure of covalently linked glycosides. Chemical bond rupture is irreversible and the decomposition E_a is much higher than the E_a of conformational distortions and viscous flow.

Raemy and Schweizer (1983) assert the following thermochemical stabilities:

(i) 1,4-β-glycans are more stable than 1,4-α- and 1,6-α-glycans
(ii) glucopyranoses are more stable than fructofuranoses
(iii) uronic acids are among the least stable polysaccharides

Specifically, D-glucose < maltose < maltotriose < amylose < starch < amylopectin < cellulose (Greenwood, 1967). Trends indicated are that thermochemical stability increases with the DP, branching, and 1,4-β bonding. Chemical bonds other than 1,4-α and 1,4-β introduce heat and acid instability. Either of these two bonds is less easily depolymerized when the sixth pyranose carbon is oxidized to the carboxyl group rather than esterified; for this reason, low-methoxyl pectin is more stable than high-methoxyl pectin.

III. Low-Temperature Pyrolysis

Pyrolysis (Irwin, 1979, 1982; Tomasik *et al.*, 1989b) is the decomposition of a substance at elevated temperatures, principally by dry heat. Low-temperature pyrolysis arbitrarily refers to thermochemical decomposition in the 121.1–

300°C range, where polysaccharides are moderately stable for short time intervals, but, after prolonged heating, there is an initial chemical dehydration, followed by chemical bond rupture that produces lower DP fractions, acids, maltose, isomaltose, glucose, and an increasingly complex vapor phase. Acidity in the involatile phase is an exponential function of time at a constant temperature (Walter and Fagerson, 1970); simultaneously, there is an accumulation of levoglucosan (1,6-anhydro-β-D-glucose). The decomposition is initially autocatalytic, with a rate increasing from lower to higher temperatures. Metal ions, H_3O^+, and OH^- catalyze the decomposition. Spectrally, the pyrolysate shifts from an ultraviolet-nonabsorbing species to an ultraviolet-absorbing chromophore. The pyrolytic pathways and pyrolysate properties, to varying degrees, are dictated by the substrate, its purity, moisture content, the heating medium, and the time–temperature integral. The most striking chemistry of low-temperature pyrolysis involves condensation, addition, transglycosylation, and the construction of branched polymers in the involatile phase; the reactions begin in the amorphous regions (Major, 1958). High time–temperature integrals generate furans, but in the presence of proteins the vapor-phase composition shifts to pyrazines.

The low-temperature pyrolysis of acidified starch (Horton, 1965; Greenwood, 1967) for a relatively short interval (4–8 h) produces "white dextrins" that are quite similar in appearance and function to unheated starch except that their reducing power and water dispersibility are enhanced, and viscosity and the retrogradation tendency decrease. "Yellow dextrins," having even lower viscosity, are produced over a larger time–temperature integral. "British gums" are the product of heating in alkali at 130–220°C for 10–20 h, coinciding with a loss of birefringence and crystallinity beginning at approximately 180°C. British gums have greater sol stability and intrinsic viscosity than white and yellow dextrins. Up to 300°C, air oxidation does not appear to be significant (Tomasik *et al.*, 1989b). Cellulose reacts similarly to starch at a faster rate in air than in nitrogen (Shafizadeh, 1968).

IV. High-Temperature Pyrolysis

Polysaccharide pyrolysis at 375–520°C is accompanied by a higher rate of weight loss and evolution of a complex mixture of vapor-phase compounds preponderantly of H_2O, CO, CO_2, levoglucosan, furans, lactones, and phenols (Shafizadeh, 1968). The volatile and involatile phase compositions are conditional on the rate of removal of the vapor phase from the heated chamber (Irwin, 1979), inasmuch as the primary decomposition products are themselves secondary reactants. The reaction kinetics is described as pseudo zero order (Tang and Neill, 1964) and zero order initially, followed by pseudo first order and first order (Lipska and Parker, 1966), suggesting an early independence of concentration, but depending terminally on one or more decomposition products of the primary reaction.

The flammability of the vapor phase of cellulose apparels gave special urgency to a search for ways to lower the combustion temperatures (lower E_a) and increase the rate of weight loss by using flame retardants, thereby rapidly augmenting the rapid accumulation of nonflammable gases at the expense of combustible distillate (Shafizadeh, 1968).

V. Maillard, Amadori, and Strecker Degradations

The chemical reactions in a solution or dispersion of a reducing sugar and an amino or imino compound (ammonia, amino acid, polypeptide, protein), aided by heat (known trivially as the Maillard reaction and scientifically as glycosylation), is one form of nonenzymatic browning of carbohydrates. Reducing carbohydrates easily oxidize to carbonyl compounds, and carbonyl compounds easily develop complexes with nitrogen compounds (Hodge, 1953). The Amadori rearrangement and dehydration multiply unsaturation in the glycosamines into brown pigments called melanoidins; these conjugates are antinutritional, because they make essential amino acids metabolically unavailable. The conjugates proceed to decomposition (Strecker degradation) and subsequently to evolution of aroma and flavor compounds, notably the pyrroles and pyrazines, from roasted peanuts, popcorn, chocolate, cocoa, cooked potatoes, beer, and bread (Belitz and Grosch, 1987), for example. Some pyrroles are constituents of coal tar—a very carcinogenic distillate of bituminous coal. The chemical composition of the pyrolysate varies with nitrogen source; each Maillard system produces a different composition in the involatile and volatile phases.

A starch–glycine mixture, heated at 290°C, was significantly different from the starch control in its composition of alkoxyphenols and imidazoles in the involatile phase, and pyrazine, pyridine, methylpyridine, and dimethylpyrroles in the volatile phase (Umano and Shibamoto, 1984). The neurotoxin 4(5)-methylimidazole appeared in the vapor phase when ammonia, but not amino acids, was the nitrogen source (Tomasik *et al.*, 1989b). A corn starch–sucrose combination inhibited the Maillard reaction (Lee and Woo, 1988).

VI. Caramels

Industrial caramel, known from the beginning to be essentially a mixture of burnt sugar polymers (Salamon, 1900), arises from the controlled action of heat on dry sugar or concentrated sugar solutions with or without acid, alkali, ammonia, or an amino compound (Tomasik *et al.*, 1989a). Polysaccharides undergo identical reactions after the initial pyrolytic decomposition. This nonspecific high DP, bitter-tasting, yellow-to-brown class of decomposition compounds is manufactured for artificially coloring food and beverages (distilled liquors, beer, carbonated beverages, soups, and candies, etc.).

The contemporaneous chemistry of caramel, not yet completely elucidated, may be summarized as follows: its variable composition is a function of time and temperature; browning first appears at approximately 160°C, whereupon the color becomes progressively darker simultaneously with frothing and evolution of a large quantity of water, traces of acetic acid, and a distillate containing furfural (2-furaldehyde); frothing ceases at about 250°C when the pyranose ring structure is completely chemically dehydrated; the residue (caramel) is soluble in water during the initial stages of decomposition, but is increasingly insoluble in direct proportion to the disappearance of the sugar; hygroscopicity and ethanol solubility increase with residual glucose content; when carbonate or ammonia is added prior to heating, the substrate develops the brown color immediately; many consecutive and competitive reactions (Houminer, 1973) yield a vapor phase of diverse compounds (Bryce and Greenwood, 1966; Walter and Fagerson, 1968). At 300°C in a nitrogen atmosphere, the dominant intermediate product was found to be dianhydroglucopyranose (Heyns *et al.*, 1966). One caramel compound was reported to approximate the formula $C_{24}H_{26}O_{13}$, reaching $C_{125}H_{188}O_{80}$ and higher (Tomasik *et al.*, 1989a).

Caramel is unintentionally generated in burnt carbohydrate foods (rice, oatmeal, cornmeal, etc.) and molasses (Kowkabany *et al.*, 1953); it is the source of maple flavor and color in the concentration of maple sap to maple syrup (Stinson and Willits, 1965). In industrial manufacturing, the intended application is taken into account, because reaction conditions help determine the properties of the pyrolysate, e.g., its tinctorial value, water solubility, and alcohol stability. Tinctorial value refers to the absorbance at 560 nm of a 0.1-wt/vol% solution in a 1-cm cell. Tinctorial strength increases with acidity, temperature, and duration of heating. Caramel manufactured above pH 6.3 is biologically unstable and much below pH 3.1, it is a resin.

VII. Summary

Throughout the range of heat processing temperatures, polysaccharides transmute and partially decompose to palatable food ingredients, appealing flavor, aroma, and color additives, leaving resins as an involatile residue. The transmuted matrix is generally recognized as safe for human consumption, although traces of antinutritional and toxic compounds may be generated.

CHAPTER **7**

Isolation, Purification, and Characterization

I. Introduction

Polysaccharides are physically and chemically characterized in attempts to correlate their structure, properties, and function. Inasmuch as they cohabit space *in vivo* alongside and interspersed with numerous biochemicals that can be expected to be extracted with them, they must be purified prior to characterization.

II. Extraction and Purification

As objects of study, polysaccharides must first be isolated from the vegetative milieu and concentrated in as high a concentration as possible, without serious structural modifications. Protein, lignin, lipid, ash, and shorter-chain congeners, for example, are invariably present in substantial or trace quantities even after rigorous purification. Amino acids and protein are particularly noticeable in commercial samples of a number of gums (Anderson *et al.*, 1986). Absolute purity is seldom achieved; relative purity is enhanced naturally by large granules (in starch) and concentrated deposits (in microbial excretions).

With exceptions, polysaccharides are customarily separated from vegetable matter with the use of aqueous reactive solvents (hot dilute acids, alkalis, oxidants, etc.). In the absence of careful control, modifications of the native structure usually attend isolation and purification with these reagents; the extent of modification is proportional to the severity and duration of the exposure. Mineral acids hydrolyze all polysaccharides, given enough strength and an adequate time–temperature integral. Dilute alkali may simply swell polysaccharide fibrils with only minor molecular changes (Jayme and Lang,

1963); strong alkali isomerizes, enolizes, and hydrolyzes glycopyranoses. Starch, for example, is decomposed after enediol isomerization antecedent to excision of simple organic acids (Schoch, 1964). High temperatures expectedly accelerate the processes.

Starch may be extracted simply with hot water, but the mildness of this exercise requires separate inactivation of depolymerases (by heat or ethanol) if this biopolymer is to be used for food, or mercuric chloride (0.01 M) if it is to be used other than for food. The crude water extracts are centrifuged, filtered, and spray- or freeze-dried in preparation for storage. Where there is protein, in small samples, most of it can be removed by shaking the crude extract with one-tenth its volume of toluene and discarding the toluene layer. Lipid matter is dissolved by refluxing with aqueous 80% methanol.

Formal procedures for starch have been outlined (Badenhuizen, 1964; Watson, 1964; Wolf, 1964; de Willigen, 1964). In one micromethod (Pucher *et al.*, 1948), sugars were dissolved in 80% ethanol and the sugar-free tissue (50–250 mg) was heated in water to gelatinize the starch; the mixture was cooled before adding 52% perchloric acid with stirring; then the mixture was centrifuged. The steps were repeated and the combined centrifugate was decanted. To 10-mL decantate was added 5-mL 20% sodium chloride and 2-mL iodine–potassium iodide reagent (75.5-g I_2 and 7.5-g KI in 250-mL water). After mixing and standing for 20 min, the supernatant liquid was decanted. The starch–iodine precipitate was repeatedly suspended in 5-mL ethanolic sodium chloride (350-mL ethanol, 80-mL water, and 50-mL 20% aqueous sodium chloride) and diluted with water. Subsequent to final centrifugation, the washed precipitate was dispersed in 2-mL 0.25-N ethanolic sodium chloride and the mixture was shaken gently until the blue color disappeared. The liberated starch was washed, centrifuged, and redispersed in 5 mL of hot water. Amylose can be further purified by cellulose adsorption chromatography, because there is a strong affinity of cellulose adsorbents for it. The adsorbed amylose is desorbed with hot water. In yet another extraction, highly purified starch was isolated from apples with 1% cold ammonium oxalate, without disintegration of the granules (Johnston, 1956).

Methods have been outlined for cotton (Corbett, 1963), bacterial (Hestrin, 1963), and wood cellulose (Green, 1963). For the highest purification, cotton may simply be washed with hot water, dilute acid, or dilute base. Relatively pure cellulose from *Acetobacter xylinum* was harvested by filtration alone (Ring, 1982). Complex cellulose forms require rigorous procedures frequently involving delignification and bleaching (with chlorite, sulfite, and hydrogen peroxide). The solid remainder after cellulose delignification is holocellulose, from which hemicellulose may be extracted with cold alkali, leaving an insoluble fraction called α-cellulose (Ikan, 1991).

In one outline, cellulose from *Valonia ventricosa*, an alga, was boiled in excess 1% aqueous NaOH for 6 h, with a change of alkali solution after 3 h; the alkali-treated cellulose was washed with distilled water (Blackwell, *et al.*, 1977), then immersed overnight in 0.05-N HCl at room temperature. The primary structure of simple cellulose, emerging swollen but mostly intact from (cold) alkali treatment, is known as alkali cellulose (Green, 1963).

In another outline, cellulose was complexed with cuprammonium ions (Nicoll and Conaway, 1943). Lately, laboratory-scale isolation has relied on polar aprotic solvents and solvent systems, e.g., dimethylsulfoxide, pyridine, *N*,*N*-dimethylacetamide–lithium chloride, and 1-methyl-2-pyrrolidinone–lithium chloride (Baker *et al.*, 1978; McCormick and Shen, 1982; Seymour *et al.*, 1982; Arnold *et al.*, 1994). These solvents have enabled such homogeneous[17] reactions as O- and N-derivatization of cellulose and chitin (Williamson and McCormick, 1994) and selective site chlorination (Ball *et al.*, 1994). Dimethylsulfoxide was the solvent in a homogeneous reaction of cellulose and paraformaldehyde, prior to isolation of purified cellulose (Johnson *et al.*, 1975). In yet another outline, paraformaldehyde enabled superior quality extracts when the parent tissues were presoaked in this solution (Fasihuddin *et al.*, 1988).

The pectin of commerce is the acid or enzyme hydrolyzate of protopectin—the insoluble parent polymer residing in the cell walls of higher plants, primarily apples and citrus fruits. The pectin is isolated by entrainment in polymeric aluminum hydroxide produced *in situ* by neutralization of an acidic aluminum salt in solution. The pectin isolate is freed of Al(III) by washing and dialysis in an acidified ethanol–water solvent. Advanced demethylation of extracted pectin with alkali or enzymes yields a series of low DE pectins.

In isolation and purification, although water is of course the main solvent, when alcohol is included, it acts as an antidispersant of the hydrocolloidal solute. Acetone is preferable when treating ethanol-tolerant polysaccharides like cellulose derivatives. The most ethanol-tolerant polysaccharides have been precipitated from water by refrigerating a dispersion of each containing a high volume of acetone.

In this laboratory, commercial samples of polysaccharides are routinely dispersed in water, then three times the volume of ethanol is added to precipitate them. Samples are finally rinsed with 95% ethanol. For the ethanol-tolerant polysaccharides, acetone dispersions are refrigerated at 0°C until the solvent and solute phases are visibly separate.

III. Analysis

Numerous methods are available for polysaccharide analysis, based either on their chemistry, occasionally involving unique fine structures and subunits, or on their response to ambient stimuli, as chain molecules sensitive to

17. Homogeneity in this sense refers to the uniform solubility (dispersibility) in the same solvent to form a single phase.

altered environments. Most reactions are understandably nonspecific, given the sameness of physical properties and many structural features. In a number of instances, the polysaccharide quantitative and qualitative methodologies share a common reagent or principle.

A. Detection of Charge and the Zeta Potential

Neutral and ionic polysaccharides are distinguished from each other by charge on the latter, which originates from dissociation of acidic groups ($-OSO_3H$, $-COOH$, $-OPO_3H_2$), complexation with ionic ligands, or adsorption of ions. The identification of charge is predicated on the polyanion's electrical response in electrophoresis and ion exchange chromatography.

1. Electrophoresis

The migration of ionic compounds in an electrical field is called electrophoresis. Relative to the counterions, polysaccharide polyanions migrate to the positively charged pole (anode), while polycations travel in the opposite direction toward the negatively charged pole (cathode); neutral polysaccharides remain at or near the site of the initial placement. Resolution of a polysaccharide mixture exploits mass-to-charge ratio differences, differential solvent compatibilities, and ζ. As the migrating polyanion encounters viscous resistance from the solvent, an electromotive force is generated (streaming potential). The rate of migration is defined as the electrophoretic mobility Ω—the mobile equivalent of ζ [Eq. (3.7)]. A mixture of different mass-to-charge ratio polyanions that have the same or close Ω in one mobile phase may be resolved further by changing one or more of the solvent variables (solvent system, pH, i, etc.). An accurate Ω rests on an independence from particle geometry, an absence of solvent convection currents, and solute–solute interactions. With foreknowledge of the other variables, Eq. (3.7) may be adapted to measure ζ.

Pechanek *et al.* (1982) determined ionic polysaccharides by Ω migration through polyacrylamide and agarose gels and on cellulose acetate membranes; the polyanions were detected by staining. At the dimensions found in gel micropores, pairs of surfaces create an adsorption potential (ξ) 3.5 times that created at the same distance from a single surface (Vold and Vold, 1983).

In a method of capillary electrophoresis, 1 pg of dextran was dispersed in alkaline buffer containing fluorescein, and the dextran was detected by negative fluorescence (Richmond and Yeung, 1993).

2. *Anion Exchange Chromatography*

Anion exchangers retain polysaccharide polyanions on a positively charged resin (RNH_3^+) in exchange for OH^- or Cl^- counterions in proportion to their charge density; neutral polysaccharides pass through freely. The most densely charged molecules adsorb closest to the sample placement site.

Aldonic, uronic, and ascorbic acids, lactones, and N-acetylated amino sugars were separated on sulfonated polystyrene–divinylbenzene, a strong polyanion exchanger (Wheaton and Bauman, 1953). This method is adaptable to neutral carbohydrates without complexation or adsorption, by immersion in strong alkali to ionize the hydroxyl groups (ion chromatography).

3. *Conjugation*

Dickmann *et al.* (1989) developed an anion-specific method of determining food gums based on complexation with horseradish peroxidase and color development with benzidine and a protein ligand. Keijbets (1974) used a modified copper acetate–arsenomolybdate assay to distinguish between hexose and hexuronan end groups. An application to carboxyl polysaccharides of the cation–anion complexation reaction involved poly(hexamethylenebiguanidinium) chloride, without eliciting any reaction from neutral polysaccharides (Kennedy *et al.*, 1992). Using hexadecyltrimethylammonium bromide, Cui *et al.* (1993) obtained two fractions from a mustard polysaccharide—a major insoluble and a minor soluble fraction. Polysaccharide polyanions can be complexed with quaternary ammonium cations (e.g., positively charged cetylpyridium) to form differentially soluble salts (Scott, 1965).

4. *Electroviscosity*

An initial negative slope of η_i vs c_i in a dilute water dispersion (electroviscosity) of a polysaccharide is indicative of polyanionic character. Electroviscosity disappears in excess electrolyte solution and is nonexistent in neutral polymer viscosity profiles.

B. Functional Group Identity

Infrared spectroscopy (IR) exploits the absorption of infrared radiation in the 400–4000-cm^{-1} segment of the radiation spectrum. IR is a generally useful method to help elucidate organic chemical structures (Barker *et al.*, 1956), including the identification of ionizable groups. Thus, IR spectroscopy is an indirect means of detecting charge. Polysaccharides are best examined as a thin (1–1.0-nm thick) film, made by spreading a 0.05–0.5% aqueous dispersion on a clean smooth surface (e.g., a watch glass), drying the film in a stream of nitrogen, and prying loose the dried film (with a spatula). Residual moisture is brought to an absolute minimum by storing the dried film over concentrated sulfuric acid (sp gr 1.84; relative humidity less than 3%).

The fundamental carbonyl ester band stretches and bends in the 1725–1749-cm^{-1} region, with overtones in the vicinity of 3430 cm^{-1}. OH^- stretch vibrations are in the 3000–3600-cm^{-1} range. The acidic carbonyl group absorbs intensely at 1736 cm^{-1}, allowing for changes in dipole moments with different neighboring organic groups. Sulfate absorbs in the vicinity of 1240 cm^{-1}, making it distinguishable from –COOH. IR bands that are most helpful in detection and quantification of modified celluloses have been listed (Barker, 1963). Ambiguities in interpreting a polysaccharide IR spectrum are obviously possible from overlap of absorbing bands, given the similarity and proximity of many subunits in the macromolecule.

The absorption band at 3600–3000 cm^{-1} (OH stretching) can be eliminated by deuteration, which makes it possible to distinguish between OH groups in the amorphous and crystalline states in cellulose where, in the former, disordered regions facilitate the deuterium–hydrogen exchange; the crystalline regions are refractory to the exchange. The ratio of absorptions before and after deuteration can provide a measure of the degree of crystallinity (Barker *et al.*, 1956).

C. Interaction and Conformation

The 1400–800-cm^{-1} segment of the IR spectrum is sensitive to polymer conformational changes (Belton *et al.*, 1986). In this segment, Wilson *et al.* (1988) studied ion–sulfate interaction; they reported an increased ion–sulfate affinity and a higher degree of ordering during gelation of κ-carrageenan with potassium. By monitoring the 1046 cm^{-1} band, they followed the course of crystallization of starch from gelatinization to a 21-day storage condition.

Conformational changes are easily followed by optical rotation (Hui and Neukom, 1964). Circular dichroism spectroscopy (CD) of polysaccharides (Morris, 1994) exploits optical anisotropy. In a CD instrumental design, the clockwise and counterclockwise rotation of two polarized beams of equal intensity, traversing a 180° path through a chiroptical medium, display a molar ellipticity maximum and minimum. CD is the differential measurement as a function of λ. By CD spectroscopy, mixed interchain association rather than nonspecific incompatibility or exclusion was identified as the molecular basis of alginate–polyguluronate interaction (Thom *et al.*, 1982).

D. Polydispersity

Polysaccharide size polydispersity transcends decades of molecular weights (Fig. 3 in Chapter 5). Such polydispersity is evaluated by a variety of methods (Barth and Sun, 1991).

1. *Electrophoresis*

Given that Ω is a function of mass-to-charge ratio, the number of spots displayed on an electrophoregram is a semiquantitative indicator of the polymolecularity of the parent polysaccharide (Aspinall and Cottrell, 1970).

2. *Chromatography*

Partitioning by gel chromatography, customarily performed under relatively low pressures (1 atm; low-pressure liquid chromatography, abbreviated to LPLC), has been adapted to separations at much higher pressures in a method referred to as high-pressure liquid chromatography (HPLC). As in LPLC, this latter mode partitions a population of molecules at the interface between a stationary liquid and a mobile liquid, but in specially designed systems that can withstand the high pressures. HPLC enables higher resolution than LPLC.

HPLC has been adapted to industrial polysaccharides (Barth and Regnier, 1981), e.g., guar (Barth and Smith, 1981), starch (Kobayashi *et al.*, 1985), polydextrose (Thomas *et al.*, 1990), pectin (Schols *et al.*, 1989), and other anionic gums (Voragen *et al.*, 1982). Baseline separation is limited to DP = 10 (Chester and Innis, 1986), which is more the size of an oligosaccharide than a polysaccharide. Separations to DP = 30 are possible under special conditions (Praznik *et al.*, 1984).

In normal polysaccharide HPLC, the most polar molecules with identical $\overline{M}$ are the last to elute in a polar solvent, because of their greater interaction with the aqueous stationary phase. Reverse-phase HPLC is the technique of substituting the LPLC stationary phase with a nonpolar solvent stationary phase and using a less polar mobile phase, with the result that the most polar homologs elute first. Heyraud and Rinaudo (1991) applied reverse-phase HPLC to the separation of low DP anomeric dextrins and Voragen *et al.* (1982) applied it to analysis of pectin enzyme digests.

Thin-layer chromatography (TLC) is another liquid–liquid partition technique applicable to polysaccharides, but in two dimensions. In TLC, the $\overline{M}$ cutoff boundaries between separated molecules are sharpened, because diffusion is minimized or eliminated in favor of capillary transport. The sample capacity of a TLC plate is in microliters. Resolution is enhanced further at high solvent pressure (Rombouts and Thibault, 1986).

Field-flow fractionation is a chromatographic method of separating components in a dispersion or suspension traveling parabolically through a narrow, empty channel in which the carrier stream is subjected to a perpendicular force field (thermal, electrical, centrifugal, etc.). Partitioning is based on molecular size, mass, charge, and density, in response to the force field (Giddings *et al.*, 1980; Barth and Sun, 1991). The residence time in streamline flow is shortest for the smallest molecules and particles; i.e., the smallest molecules and particles are the first to elute. The slow response time of the largest molecules and particles enables them to migrate during flow toward the wall of the tube where the velocity is lowest, which therefore makes them the last to elute. Field-flow fractionation has been touted as a speedy, high resolution technique.

By applying a steric subtechnique (Giddings *et al.*, 1980), field-flow fractionation was shown to have the potential to separate stable polysaccharide suprastructures in greater than hydrocolloidal diameters. Moon and Giddings (1993) used the procedure to size starch granules into a bimodal distribution of mass greater and less than 10 μm.

3. Miscellaneous Methods

Broad categories of starch granule sizes are possible by fractionation with butanol (Schoch, 1942). This solvent enters the interior of the amylose helix and forms an insoluble inclusion complex.

Photon correlators measure scattered light in a sol, equating this with size, and particle counters measure the conductivity or capacitance of dispersed solute; calibration is necessary. From particle diameters, volumes can be calculated, assuming a spherical geometry. Sizes and distribution are reported as histograms (Fig. 1).

By laser diffractometry, Okechukwu and Rao (1996a) found that ungelatinized cowpea starch granules had a unimodal distribution with a mean of 19 μm. In an unrelated method, Chuma *et al.* (1982) used photography, a digitizer, and a microcomputer to calculate the size, surface area, and volume of grains and soybeans.

IV. Molecular Weights and Sizes

Molecular weights are determined by end-group analysis ($\overline{M}_n$), membrane osmometry ($\overline{M}_n$), viscometry ($\overline{M}_v$), size exclusion chromatography ($\overline{M}_w$), light scattering photometry, and sedimentation ($\overline{M}_w$). Any molar mass computed by these methods must be evaluated critically, in view of a dependence on methodology.

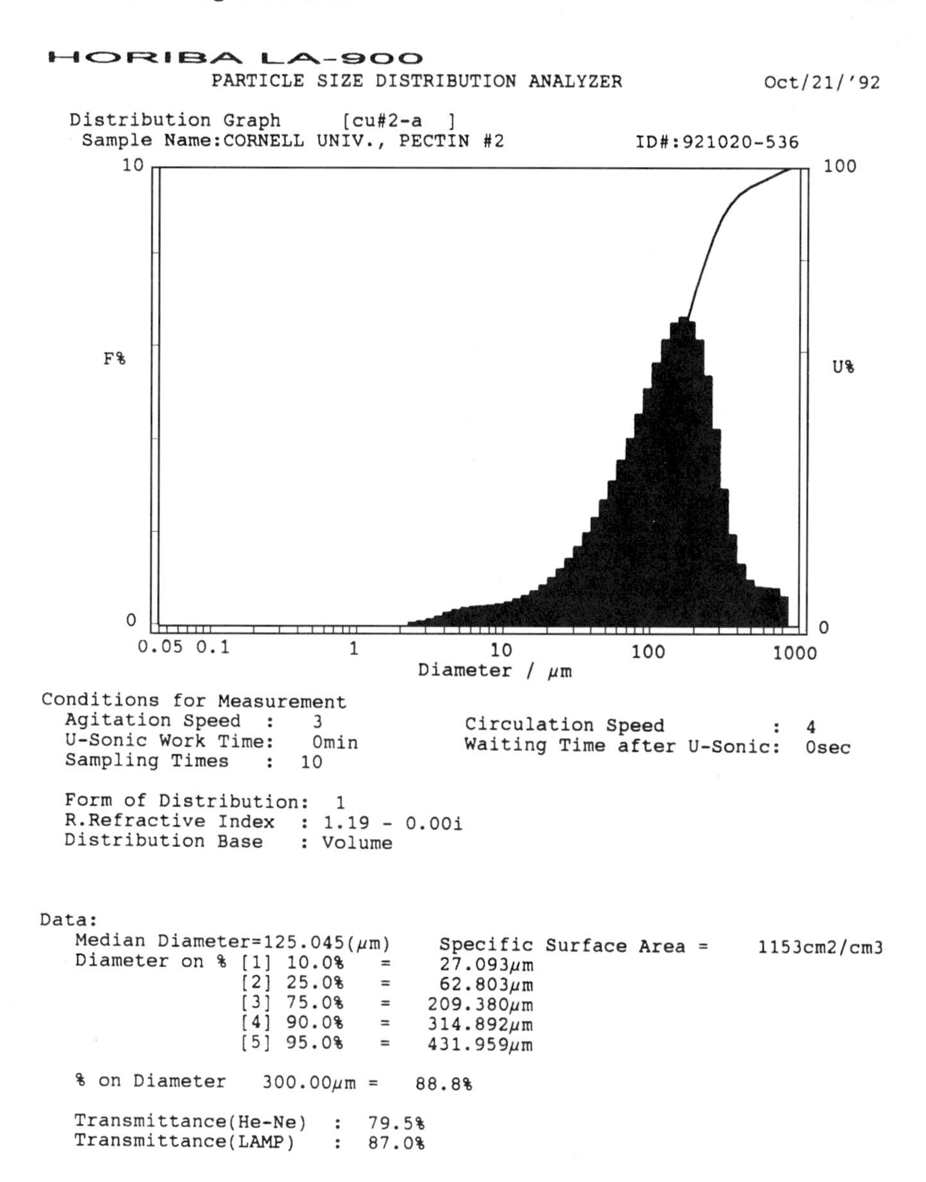

Figure 1 Size distribution of a sample of pectin (Hercules, Inc., Wilmington, DE). Courtesy of Horiba Instruments Inc., Irvine, CA.

A. Reducing End-Group Analysis

Reducing end-group analysis (Smith and Montgomery, 1956) is implicitly the most direct way to determine molecular weights of linear polysaccharides, because a chemical reaction establishes one-to-one correspondence between each polymer molecule and a reagent molecule, with the result that the measurement is theoretically indifferent to mass and polydispersity, relying solely on a theoretical, exact, but from practical experience, not necessarily stoichiometric quantitation between the two reactants. There is a DP sensitivity limit of approximately 2.5×10^4 Da (Garmon, 1975), suggesting polysaccharide oligomers to be most amenable to this method. Comparison of the results with other methods is good to poor (Launer and Tomimatsu, 1959). Branched polysaccharides do not present one-to-one correspondence, consequently making reducing end-group analysis unsuitable for amylopectin and glycogen.

The molar ratio of polysaccharide reducing end groups to nonreducing end groups can be determined alternatively by methylation (Ingle and Whistler, 1964) and periodate oxidation (Mehltretter, 1964; Shasha and Whistler, 1964; Whelan, 1964). A common procedure for starch is oxidation to the dialdehyde, whereby each mole of nonreducing and reducing end group yields a definite number of moles of formaldehyde and formic acid. From the data, a ratio of nonterminal to terminal glucose monomers can establish the DP, $\overline{M}_n$, and branching. DP heterogeneity results in wide deviations from the mean.

B. Viscometry and Rheometry

Single-point viscometers allow one η_{sp}/c_i measurement at a time. Dilution viscometers allow for continual dilution in the cell by adding solvent calculated from the equation

$$V_1 c_1 = (V_1 + v) c_2 . \quad (7.1)$$

V_1 is the volume of sample at concentration c_1 and v is the volume of solvent to be added to accomplish the next dilution to c_2. Nine or ten dilutions should be made, but no less than three, so that η_{sp}/c_i vs c_i contains a minimum of four points, excluding $c_i = 0$ where extrapolation yields $[\eta]$. In either single-point or dilution viscometry, the basic η_{rel} is measured first, precedent to the secondary equations in Table I in Chapter 4. Having plotted and extrapolated η_{sp}/c_i vs c_i to $[\eta]$ and knowing K and ν beforehand, substitutions may be made in the Mark–Houwink equation [Eq. (4.47)]; $\overline{M}_v$ can thus be calculated.

By coordinate orientation, small $[\eta]$ differences are exaggerated by the exponent in Eq. (4.7), making coordinate orientation suitable for evaluating functionality and making possible a studied selection for a particular use. If, for example, a polysaccharide giving a low viscosity in an ethanol solution is required, C or SS(2) is the pectin of choice (Fig. 2). The response to ethanol is shown to be pectin-specific rather than class-specific. Instead of ethanol concentration, the abscissa may be c_i (Walter, 1991).

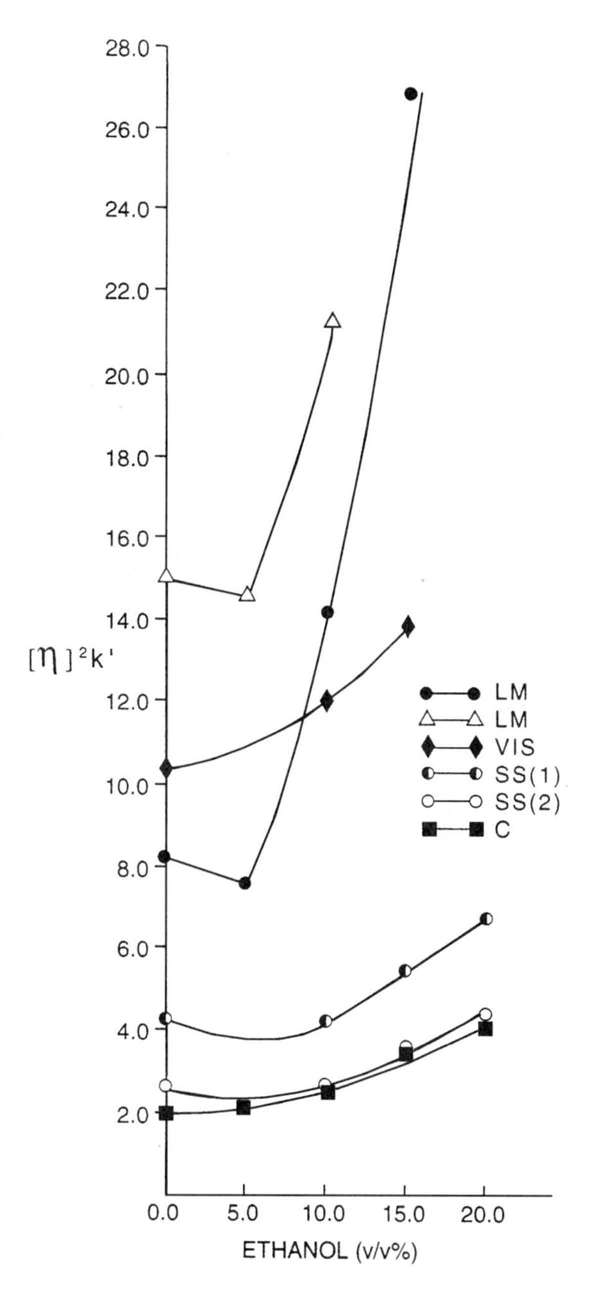

Figure 2 Viscosity profile of selected samples of pectin in aqueous dispersions, as a function of ethanol content.

A rheometer measures higher dispersion concentrations than a viscometer and, unlike the latter, can also measure suspensions.

C. Size Exclusion Chromatography

The first fraction of solute emerging from a gel column is eluted in v_0 that contains molecules too large to enter the gel micropores. v_0 is the macroscopic pore space in the gel bed, not otherwise participate in the sieving mechanism. For a given column, v_0 is constant. The solute interacting with the liquid stationary phase on the column surface elutes in order of the magnitude of a fraction's partition coefficient (K_p) between the elution volume (v_{el}) and the volume of stationary solvent in the micropores (v_s), fixed at 100 mL, because of difficulty in its measurement (Bio-Rad, 1971). At $V_s = 100$ mL,

$$K_p = (v_{el} - v_0)/(100 - v_0). \tag{7.2}$$

It is self-evident from Eq. (7.2) that the smallest sizes possess the largest K_p, because they are the last to emerge from the column (largest v_{el}). K_p is relatively large also, if there is bonding between the solid support and the eluting solute, for the obvious reason that bonding extends the retention time and hence v_{el}. K_p is variable with temperature.

The steepness of the slope of an ideal straight line K_p vs log M is an index of the resolution efficiency of a gel bed: the steeper the slope, the higher is the baseline resolution. If K_p is plotted against the logarithm of standard $\overline{M}$ of a homologous series, an unknown $\overline{M}$ should ordinarily be determined from the plot, but for the difficulty in procuring a homologous standard series and the susceptibility of K_p to deviations arising from interaction (Rollings *et al.*, 1983; Anger and Berth, 1985), charge, and polymolecularity.

The difficulties presented by log K_p vs log of standard M are obviated by alternatively plotting log[η]M vs log v_{el}. For each fraction of a polysaccharide homologous series, log[η]M vs log v_{el} is superimposable on a so-called universal calibration curve (Grubisic *et al.*, 1967) from which an unknown M_2 may be estimated, after substitution in the equation

$$[\eta]_1 M_1 = [\eta]_2 M_2. \tag{7.3}$$

$[\eta]_1$ and M_1 are values of a standard, e.g., dextran, and $[\eta]_2$ and M_2 are values of the unknown. Berth and Lexow (1991) measured pectin M by using Eq. (7.3). Barth (1986) derived Eq. (7.4) by combining ν and K of the Mark–Houwink equation [Eq. (4.47)] with Eq. (7.3) (Appendix 6):

$$\log M_2 = [(\nu_1 + 1)\log M_1 + \log K_1 - \log K_2]/[\nu_2 + 1]. \quad (7.4)$$

Size exclusion chromatography did not differentiate lower molecular weight, extended coils from higher molecular weight, compact coils (Berth, 1988).

D. Membrane Osmometry

In membrane osmometry, use is made of the fact that the chemical potentials of water (μ_o) and of dispersion (μ_i), separated by a membrane permeable to water only, are unequal except at equilibrium, and that π (a function of $\Delta\mu$) is directly proportional to $\overline{M}_n$. Accurate measurements are confined to a practical upper limit of 10^5–10^6 Da (Garmon, 1975; Dautzenberg *et al.*, 1994).

Membrane osmometry is executed in a static or dynamic mode. During static osmometry, preequilibrium aging may decrease the number and increase the average size of polysaccharide particles. Dynamic osmometry rids the measurement of aging as a source of error, but the small volumes of sample used are prone to air saturation that distort the π readings. Aggregation lessens the number of colloidal units and consequently lowers $\overline{M}_n$; disaggregation has the opposite effect. Semipermeable membranes present problems arising from flexibility of the membrane, from outward migration of molecules in the lower colloidal range, and from circumferential solvent diffusion. The starting electrolyte concentration should be equal on both sides of the membrane to avoid initial pressure surges. The "ballooning" of flexible membranes gives a false reading of c_i [i.e., g/V_i; Eq. (4.29)]. The outward migration of the smallest particles across the membrane artificially decreases the $\overline{M}_n$ and narrows the distribution range. Circumferential solvent diffusion results in loss of an increment of π in proportion to the weight of escaped water.

Donnan distribution, leading to abnormally high π, is a most serious source of error for $\overline{M}_n$ of polyanions measured by osmometry. The Donnan effect is allegedly overcome by very dilute concentrations that never exceed 25 g L^{-1} for the polyanion and $i = 0.3$ mol L^{-1} for the solvent (Wagner, 1949). A 25-g L^{-1} polysaccharide concentration is in the semidilute-to-concentrated domain, outside the theoretical dilution limit of osmometry where a power-law dependence of π/c_i on c_i is expected.

Equations (7.5)–(7.11) illustrate the effect of Donnan distribution on osmometry. Recalling Eq. (4.29) ($\pi V_i = n_i RT$), assuming an equivalent weight of an ionic polymer (i) is in equilibrium in water with three equivalent counterions (1, 2, and 3), and inasmuch as π is a colligative property, the number of dispersed particles $\mathbf{N}n_i$ is

$$\mathbf{N}n_i = \mathbf{N}c_i/\overline{M}_n, \tag{7.5}$$

$$\mathbf{N}n = \mathbf{N}(n_i + n_1 + n_2 + n_3), \tag{7.6}$$

$$\pi = (\mathbf{N}RT/V_i)\left(c_i/\overline{M}_n + c_1/M_1 + c_2/M_2 + c_3/M_3\right). \tag{7.7}$$

If the counterion is H^+ originating from dissociation,

$$\pi = [\mathbf{N}RT/V_i]\left[\left(c_i/\overline{M}_n\right) + 3c_1/1\right]. \tag{7.8}$$

If the counterion is Na^+, letting ς be the equivalent of Na^+Cl^- diffusing to the polyanion (P^{3-}) from an outer volume at y molar concentration, and $\not{c}$ be the equivalent of H^+ not diffusing outward, the equilibrium condition is

$$P^{-3} + \varsigma Na^+ + \varsigma Cl^- + \not{c}H^+ \rightarrow (y - \varsigma)Na^+ + (y - \varsigma)Cl^- + (3 - \not{c})H^+ \tag{7.9}$$

and for $\overline{M}_n$,

$$\overline{M}_n = c_i\mathbf{N}RT/(\pi V_i) + \mathbf{N}RT/(\pi V_i)[(\not{c}/1) + (\varsigma/23) + (\varsigma/35)] \tag{7.10}$$

$$= (\mathbf{N}RT/\pi V_i)(c_i + \not{c} + \varsigma/23 + \varsigma/35). \tag{7.11}$$

Relative to $c_i/\overline{M}_n$, $\not{c}$, $\varsigma/23$ and $\varsigma/35$ are high molar contributions to $\overline{M}_n$.

Notwithstanding the problems associated with the Donnan distribution, a pectin $\overline{M}_n$ was obtained from $c_i = 10^{-2}$ g mL^{-1} dispersion in 0.05-M sodium chloride and reported to have approximated $\overline{M}_n$ by reducing end-group analysis (Fishman *et al.*, 1986).

An osmometry $\overline{M}_n$ of a standard and an unknown polysaccharide in the same solvent system may be substituted with [η] in the Mark–Houwink equation, and K and ν may be calculated from two simultaneous equations.

E. Light-Scattering Photometry

Light-scattering analytical methodology is plagued by the effect of extraneous particles, leading to erroneously large chain dimensions (Jordan and Brant, 1978). In measuring I and I_ψ, the sample must therefore first be freed of extraneous particles (dust, fiber, globules, etc.); this is accomplished by terminal filtration directly into the measuring cell through a 0.45-nm filter (Berth, 1992). Approximately 5% of a pectin solute was discovered in this laboratory to be retained by a 0.45-nm hydrophilic acrodisc (Gelman Sciences, Ann Arbor, MI 48106) and approximately the same quantity to be lost by ultracentrifugation. There is the possibility that spontaneous aggregation in a polysaccharide sol may be the source of the fraction removed. A polysaccharide light-scattering $\overline{M}$ can differ from a reducing end-group $\overline{M}$ by more than threefold (Veis and Eggenberger, 1954) and an osmometry $\overline{M}$ by more than twofold (Walter and Matias, 1991).

Dynamic light scattering, coupled with modern computer programs and auxiliary equipment, automatically graphs Zimm plots and computes $\overline{M}_w$, R_g, ξ, Ω, and flow rates. New techniques have expanded the method to the semidilute and concentrated regimes (Barth and Sun, 1991).

F. Sedimentation Equilibrium and Sedimentation Velocity

In a sedimentation equilibrium experiment, a straight line for Eq. (4.76) is indicative of size homogeneity; curvature is in relative proportion to the degree of molecular weight and size heterogeneity. The experimentation time is considerably shortened by inserting c_0 measured at the meniscus (x_0) and c_1 measured at the bottom of the cell (x_1), assuming that the concentration gradient at these two locations is invariant after a short interval (Scholte, 1975).

Horton *et al.* (1991) measured $\overline{M}_w$ of sodium alginate by sedimentation equilibrium and showed that an accurate determination required inclusion of A_3 [2β in Eq. (4.8)], even at very low concentrations (2.5 mg mL^{-1}). In one of the very few instances in which sedimentation velocity was applied to polysaccharides, Sharman *et al.* (1978) determined $\overline{M}_w$ of several galactomannans in different mannose:galactose ratios and found an independence of the Mark–Houwink equation from this polydispersity. From sedimentation data, in conjunction with data from other methodologies, Harding *et al.* (1991b) calculated the mass per unit length of citrus pectin fractions approximating 430 g mol^{-1} nm^{-1}, which suggested a rod or wormlike coil with a long persistence length. Wedlock *et al.* (1986) found good agreement between a sedimentation and a laser light-scattering $\overline{M}_w$.

V. Colorimetry and Spectrophotometry

Unsubstituted polysaccharides do not appreciably absorb ultraviolet and visible radiation, but they can be made to do so intensely by combining them with chromophores and chromogens (e.g., α-naphthol, dihydroxynaphthalein, anthrone, carbazole, phenol-sulfuric acid, 2-thiobarbituric acid, toluidine blue, diphenylamine, Congo red, aniline blue, and methyl orange), usually in acidic or basic media. Coloration is normally preceded by depolymerization, deesterification, enolization, and dehydration to oligomers, shorter-chain hydrolysates, anhydro rings, and double bonds.

There are color reactions for general classes of carbohydrates (Dische, 1962), e.g., those in Table I summarize a few general-purpose tests for routine differential staining of plant-tissue isolates.

Chlorozinc-iodine, made by adding a solution of 10 g KI and 0.15 g I_2 in 10 mL water to 90–100 mL of a 60% $ZnCl_2$, can identify lignin in a plant extract by its yellow-to-brown color. Cellulose is distinguished from noncellulose matter by its blue-to-violet color (Greenish, 1923). Starch interferes with the reaction, similarly turning blue-to-violet. The color disappears at elevated temperatures and reappears upon cooling. Uranyl acetate is a negative stain for crystalline cellulose; observed under an electron microscope, it shows crystallites inhabiting the translucent areas surrounded by stained amorphous cellulose (Heyn, 1966).

Polysaccharides develop color with alkaline hydroxylamine (NH_2OH) and Fe^{3+}. The reagent is made by combining 4 g NH_2OH, 14 g NaOH, and 10 g $FeCl_3$ in 100 mL 0.1 N HCl diluted 1:3. The alkali deesterifies the polysaccharide, and the deesterified molecules then develop color with Fe(III) (Doesburg, 1965; McReady and Reeve, 1955; Bean and Bornman, 1973).

Starch in helical conformation is indicated by the blue color developed with iodine. In a typical test, a tissue or extract is submerged in a KI_3 solution. DP heterogeneity is a factor; the higher the DP, the higher is the I_2 absorption and the more intense is the violet–blue color. The starch–iodine reaction is sensitive enough for starch to be a titrimetric indicator of I_2, mindful of the nonstoichiometry and nonspecificity of the reaction.

The I_2–starch reaction is the basis of a spectrophotometric assay at 640 nm with amylose standards. A similar Congo red method is less precise than the iodine method, but it has the advantage of insensitivity to the DP, molecular size, and shape. The Congo red assay can therefore be supple-

TABLE I
Color Reagents and Tests for Some Polysaccharides

Test	Cellulose	Lignin	Starch	Pectin
Chlorozinc-iodine	b–r–v	y–b		
NH_2OH-$FeCl_3$				p
I_2			b	
I_2-H_2SO_4	b	y–b		
Carbazole-H_2SO_4				r
Ruthenium red	g	g		p
Phloroglucinol		r		

b, blue; r, red; v, violet; y, yellow; p, pink; g, gray.

mentary to the iodine assay, to minimize errors caused by heterogeneity (Carroll and Cheung, 1964). The amount of I_2 complexing with amylopectin (and glycogen) is much less than with amylose, and the I_2–amylopectin complex is red-to-brown (Kerr, 1950). The intensity of the different colors with amylose and amylopectin can be used to differentiate waxy starch from ordinary starch.

Of a series of polysaccharides tested, xanthan gum was the only one that reacted positively to toluidine blue and methylene blue (Nakanishi *et al.*, 1974); the latter dye reacts with polysaccharide polyanions including carrageenans, with which it develops a purple color.

Quantitative spectrophotometric methods for pectin utilize carbazole (diphenyleneimine; Bitter and Muir, 1962) and *m*-hydroxydiphenyl (Kintner and Van Buren, 1982). The intense red-to-brown color with carbazole in sulfuric acid, relatively specific for uronans (pectin and alginate), is much less intense with ketohexoses, aldohexoses, and pentoses (Snell and Snell, 1953). The *m*-hydroxydiphenyl assay is subject to less interference than the carbazole assay.

Ruthenium red (ammoniated ruthenium oxychloride) is a strong indicator of the polycarboxylic acid groups in pectin. This reagent is made by adding enough ruthenium red powder to 10% lead acetate to produce a wine-red color.

Starch, pectin, cellulose, and cellulose derivatives are assayed with anthrone (Viles and Silverman, 1949; Samsel and Aldrich, 1957), and methylcellulose is assayed with diphenylamine (Kanzaki and Berger, 1959). 2,7-Dihydroxynaphthalein (2,7-naphthalenediol) develops a blue–red color with glycolic acid abstracted from the carboxymethyl group of acidified CMC (Graham, 1971; Harris *et al.*, 1995). Allen *et al.* (1982) developed a specific test for 3,6-anhydrogalactose in κ-carrageenan in a mixture of food gums by using 2-thiobarbituric acid. The Maillard reaction with cysteine and methylpentoses (Dische and Shettles, 1948) has been resurrected by Baird and Smith (1989a) and Graham (1990) in a specific assay for gellan in which there is 6-deoxyhexose (rhamnose). Aniline blue, reported to be a specific stain for 1,3-β-D-glucans, has been adapted to the quantitative analysis of gums containing this structure (Nakanishi *et al.*, 1974). The SO_4^{2-} ion stripped from carrageenan was precipitated with Ba^{2+} in a gravimetric method of carrageenan analysis (Hansen and Whitney, 1960). Comprehensive methodologies for a large number of other food gums have been outlined (Smith and Montgomery, 1959; Glicksman, 1969; Graham, 1977).

Newer assays involve the action of enzymes, e.g., the assay of Baird and Smith (1989b) who treated galactomannans with galactose oxidase to generate H_2O_2 that in turn oxidized *o*-tolidine in the presence of preoxidase. These authors claimed this double oxidation reaction to be specific for the galactosyl monomer and its derivatives, based on the exclusive oxidation at the C-6 position. *o*-Phenylenediamine can be substituted for *o*-tolidine; the color is measured at 425 nm.

In the current AOAC (1990) method of starch analysis, samples are ground and freed of simple sugars with hot aqueous 80% ethanol, extracted with boiling water, and subjected to the action of glucoamylase for conversion to glucose prior to addition of *o*-dianisidine; color is measured at 540 nm against glucose standards. Ethanol is a glucoamylase inhibitor, so all traces must be removed. A two-enzyme procedure for the cereal grains, with glucoamylase and glucose oxidase in sequence, has been assigned first-action status. In this case, color is developed with *o*-dianisidine; previously, color development was by ferricyanide reduction (AOAC, 1980).

Enzymes have been used to differentiate broad categories of polysaccharides. Peroxidase differentiates polyanions from neutral molecules on the principle of complexation of the polyanion with a protein cation and a positive reaction to a protein-specific stain (Dickmann *et al.*, 1989). An immunoassay for the detection of galactomannans (locust bean and guar gums) in the range of 10 ng mL^{-1} was developed by Patel and Hawes (1988). The gums were captured on an immobilized lectin that specifically bound galactose; a peroxidase oxidized the substrate to a chromogen and measurement was made at 490 nm. Some enzymes identify specific structures, e.g., in the microassay of Ostgaard (1992) for alginate (0.01–1 mg mL^{-1}) in which the molecule was split at the nonreducing end with a lyase and the concomitant unsaturation was measured by ultraviolet absorption at 230 nm.

VI. CD and NMR Spectroscopy

The circular dichroism (CD) spectrum of a number of polysaccharides is close to that of the corresponding monosaccharides (Morris, 1994). CD information gleaned from monosaccharides and extrapolated to polysaccharides is valid, insofar as mono- and oligosaccharides are representative of the complete polymer structure. CD applications to polysaccharides have been reviewed (Johnson, 1987). With the use of CD and complementary instrumentation, Morris (1976) studied xanthan insensitivity to salt and temperature, and observed that stability was introduced by a folding back of side chains around the main chain. Dentini *et al.* (1991) outlined the use of CD to measure the average charge density of pectin chains.

Nuclear magnetic resonance (NMR) spectroscopy is routinely applied to small carbohydrate molecules. NMR spectroscopy is based on the principle that radiofrequencies are absorbed by hydrogen and carbon atoms (^{1}H and ^{13}C) spinning in one of two directions (spin quantum number $+1/2$) in a magnetic field. In liquids, absorption is recorded as sharp peaks. The frequency displacement (chemical shift) is a function of the ^{1}H and ^{13}C surroundings. $+\Delta E$ is proportional to the number of photons absorbed between these two quantum states, correlating well with anomeric and conformational changes. NMR spectroscopy was primarily a liquid-state method, but newly developed techniques (Segre and Capitani, 1993) have permitted solid-state structures and dynamics to be studied in great detail. Solid-state spectra are sometimes difficult to interpret, because of peak broadening.

By NMR spectroscopy, polysaccharide structure (McIntyre and Vogel, 1993) and heterogeneity (Cleemput *et al.*, 1993) have been studied. Kasai and Harada (1979) assigned a helical conformation to heated (60°C) curdlan gels with $DP > 49$; Morgan *et al.* (1992) studied the relationship between bread staling and starch crystallization; Cooke and Gidley (1992) discovered that the double helix is more prominent in native starch than crystals and that both structures are concurrently disordered during gelatinization; Morrison *et al.* (1993) showed that lipids are present as inclusion complexes with V-amylose.

NMR spectroscopy is quite conducive to the study of water relations. Radosta *et al.* (1989) discovered that the amount of monomolecular water bound to maltodextrin was independent of state (sol or gel), but was slightly dependent on temperature.

NMR spectroscopy has been adapted to the study of relaxation, because the excited atoms return to their respective ground states at definite rates, depending on the nuclear environment (Gidley, 1992). Harris *et al.* (1995) presented a synopsis of the potential use and limitations of NMR applied to polysaccharides.

VII. Thermal Analysis

Physical and chemical events are accompanied by $\pm\Delta H$ [Eq. (3.12)] in the form of heat loss or gain ($+\Delta C_{p,v}$) that is measurable by a number of instrumental techniques. Differential thermal analysis (DTA) quantitatively relates $\Delta C_{p,v}$ to ΔT (the difference between the reference and sample temperatures). DTA suffers the handicap of being influenced by the different heat conductivity and bulk density of the sample and reference (Cowie, 1991). In differential scanning calorimetry (DSC), electrical energy from an independent reference and sample heater, necessary to maintain the reference and sample at the same temperature, is measured. Modern instrumentation and computer software make DTA and DSC virtually indistinguishable. In an experimental DTA and DSC design, a reference and a sample are heated and cooled at a constant or variable rate; calibration is necessary. The weight loss during a pyrolytic change may be measured by thermal gravimetry (TG).

Thermal transitions and the characterizing intensities are indicated by inflections on a thermogram in a positive or negative departure from the baseline of $\Delta C_{p,v}$ vs ΔT or Δt, with different magnitudes of maxima or minima whose exact horizontal position depends on such factors as concentration, granule stability (Liu and Lelièvre, 1992), and the method of sample preparation (Morita, 1956). Time is introduced by scanning, and time-dependent phenomena, e.g., transport and relaxation, make $d(\Delta H)$ the sum of partial derivatives (Fig. 3; Provder *et al.*, 1983):

$$d(\Delta H)/dt = (\partial(\Delta H)/\partial t)_T + (\partial(\Delta H)/\partial T)_t \, dT/dt. \tag{7.12}$$

dT/dt is the experimental scanning rate. The maxima $[d^2(\Delta H)/dt^2 = -0]$ and minima $[d^2(\Delta H)/dt^2 = +0]$ are in a positive or negative direction, measured from a common, interpolated, or approximated baseline (Figs. 3 and 4). The area under the curve is proportional to ΔH. Equation (7.12) is the forerunner of an equation that enables the calculation of E_a from dynamic DSC data (Provder *et al.*, 1983).

DTA, DSC, and TG have become routine for monitoring polysaccharide solid-state transformations. Examples of important applications are hydra-

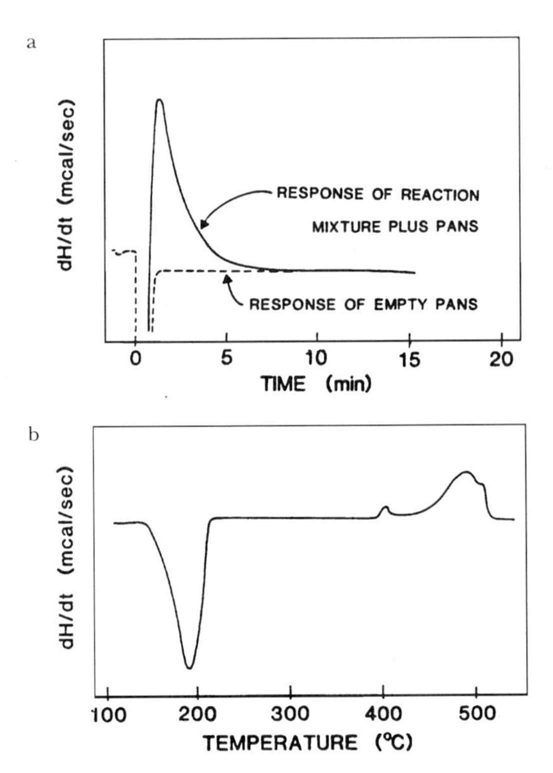

Figure 3 Characteristic DSC exotherm (a) and endotherm (b). (From Provder *et al.*, 1983. Reprinted with permission.)

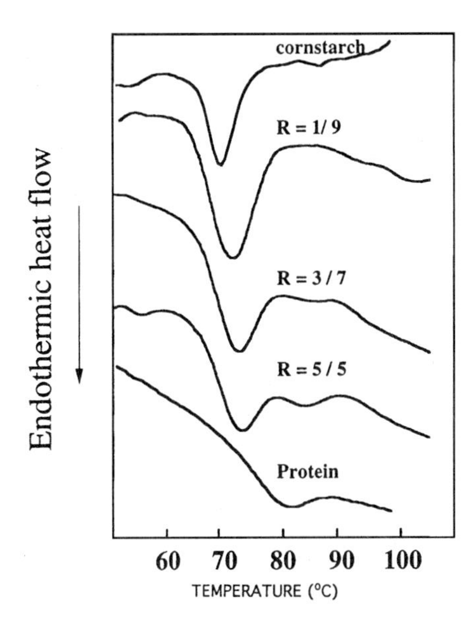

Figure 4 Differential scanning calorimetry endotherms (DSC model 2910, TA Instruments, New Castle, DE) of cowpea protein–corn starch blends in different ratios (R) at pH 7 and scanning rate 5°C min^{-1}. (Courtesy of Okechukwu and Rao, 1996b.)

tion–dehydration, solation–gelation (Krag-Anderson and Solderberg, 1992), glass transition, cold crystallization, melting, liquid crystal ordering, dehydration, rearrangement (Morita, 1956, 1957; Hatakeyama *et al.*, 1989), synergism, and antisynergism (Nishinari *et al.*, 1992), usually involving starch systems (Lelièvre, 1992).

Morita (1956, 1957) discerned characteristic endotherms in the vicinity of 130 and 255°C for polysaccharides containing predominantly 1,4-α-anhydroglucose linkages, and at 340°C for the β-linkages. Endotherms of α-anhydropolysaccharides occurred at lower temperatures than did those of β-anhydropolysaccharides. The major exothermic reaction (decomposition) occurred at 490–510°C. A specific thermal curve could not be assigned to the 1,4-α linear or the 1,6-α branched linkages. The thermograms of dried and rehydrated rice starch suggested a dehydration–rehydration cycle initiating modest irreversible conformational shifts. By TG, endothermic changes in starch at 150°C were shown to be due to loss of capillary and absorbed water and to anhydro ring formation at 200–220°C. At 270–310°C, there was sudden evolution of decomposition gases. Air oxidation seemed not to have been a factor up to 300°C (Tomasik *et al.*, 1989b). Tang and Neil (1964) demonstrated a net endothermic response of cellulose decomposed at 300°C in an inert atmosphere, and a net exothermic response in oxygen. Gekko *et al.* (1987) used DSC to show a degree of substitution (DS) effect of sugars and polyols on the gelation of κ-carrageenan. The endotherms in Fig. 4 show a slight upward shift in corn starch T_{gz}, as a result of protein additions; the protein denaturation ΔH was much lower than the starch gelatinization ΔH (Okechukwu and Rao, 1996b).

VIII. Thermodynamic Variables

DSC data are recorded on a thermogram delineated by $\Delta C_{p,v}$ vs ΔT or Δt. Inasmuch as it is the energy consumption that is recorded, the thermogram shows $+\Delta E$ proportional to the change in property. DSC is therefore a method of directly quantifying $\Delta C_{p,v}$, ΔH, ΔS [Eq. (3.16)], and $\Delta C_{p,v}/g$ (the specific heat). Exothermic reactions are characterized by positive peaks on the thermogram, and endothermic reactions are characterized by negative peaks.

IX. Structural Elucidation

Many of the foregoing analytical methods fit a strategy for elaborating polysaccharide gross structures and extracting information on intra- as well as intermolecular associations. In solid-state structural characterizations, fragments are isolated and correlations are attempted with intact structures (fingerprinting). It is not always possible to assign a particular fragmentation pattern to a specific molecule, but the distribution pattern of the fragments, generated under controlled conditions, can nevertheless provide useful information on phenomena like stability (Fang *et al.*, 1981; Liebman and Levy, 1983), complexation, and chain organization (Biliaderis, 1992). Pattern recognition is feasible, insofar as the vapor phase is rapidly flushed from the reaction chamber, so that a pyrogram, as much as possible, is a record exclusively of primary reaction products (Irwin, 1979). Polysaccharides containing the same repeating dimer, e.g., starch and cellulose, are virtually identical (Bryce and Greenwood, 1966; Sjoberg and Pyysalo, 1985) and therefore have virtually identical fragmentation printouts.

Gas chromatography (GC), also known as gas liquid chromatography, resolves mixtures of volatile compounds in a high-pressure mode in a heated gas stream, by partitioning solute between a stationary liquid phase adsorbed on a solid support and a mobile gas phase passing over the stationary phase. GC is adaptable to a pyrolysis vapor phase of simple sugars, oligo- and polysaccharides, and uronans that can be made volatile through chemical derivatization (e.g., methanolysis and silylation; Ha and Thomas, 1988).

Wheat pentosans were dehydrated to furfural with concentrated HCl, and the furfural was dissolved in dibutyl ether in preparation for GC (Folkes, 1980). This method is claimed to be specific, accurate, and precise for pentoses and pentosans (Folkes, 1980), because 5-hydroxymethylfurfural is not produced, as is the case in hexose decomposition. Previously, furfural was distilled and analyzed colorimetrically. Alkalization increases the yield of 1-, 2-, and 3-carbon, vapor-phase compounds at the expense of anhydro sugars and furans (Ponder and Richards, 1993). Hydrolysis–GC is adaptable to the estimation of DP (Morrison, 1975); applicability to carbohydrates is limited to oligomers with DP = 6 (Traitler *et al.*, 1984).

In preparative GC, each separated fragment or derivative passes through a nondestructive detector, or a minor portion of the effluent stream is diverted to the detector and the remainder collected.

The analytical advantages of pyrolysis (PY) and GC have been programmed into a sequential method of solid-state characterization of polymers. Morita *et al.* (1983) decomposed homoglucans anaerobically with heat in alkaline sodium sulfite in the presence of *o*-phenylenediamine to form the respective quinoxalines. Previously, it was necessary first to hydrolyze the homoglucans completely and then silylate the monosaccharides. Each class of glucan quinoxaline decomposed to a specific dicarbonyl compound for which the phenylenediamine reaction was specific. Hicks *et al.* (1985) offered a cation-exchange option to traditional polysaccharide anion-exchange chromatography and GC.

Supercritical fluid chromatography (SFC) is a GC method of analysis of compounds in systems where normal GC presents resolution difficulties (Lee and Markides, 1987). A supercritical fluid has properties at a critical temperature intermediate between a liquid and a gas. At and above this critical temperature, a gas cannot be compressed into a liquid, irrespective of the pressure, but it solvates solid matter as if it were a liquid. A supercritical fluid diffuses freely into and out of adsorbent pores with a minimum of resistance. A major advantage of SFC chromatography is its ability to effect separation of oligomers without derivatization.

SFC-GC and flame ionization detection (Chester, 1984) were thought of with the idea of extending the size analytical capability, while simultaneously enhancing resolution and lowering the separation temperature of silyl ethers of corn syrup carbohydrates (Chester and Innis, 1986).

Mass spectrometry (MS) is applicable to ionized polysaccharide fragments and molecular ions (Hellerqvist and Sweetman, 1990); these are generated by many techniques (Anderegg, 1990). Food gums display characteristic MS fragmentation patterns (Coates and Wilkins, 1987). Decomposition fragments can be analyzed at selected ion sites. The degree of methylation of pectin has been estimated at mass-to-charge ratios 85 and 96 (Aries *et al.*, 1988). PY-GC alone is sufficient to measure the degree of methylation of pectin (Barford *et al.*, 1986).

Advances in instrumentation have enabled qualitative and quantitative determinations of polysaccharides in the solid or liquid state by PY-GC-MS. Samples of sols may be injected directly into the gas chromatograph; xerogels may be subjected to PY or inserted directly into the ionization chamber of the mass spectrometer. The mass thermograms and spectrograms permit identification (Donnelly *et al.*, 1980; Morita *et al.*, 1983; Hellerqvist and Sweetman, 1990) through patterns coincident with known molecules: for example, an acetic anhydride pattern indicates the presence of pectin or gum tragacanth (Sjoberg and Pyysalo, 1985); methoxyphenols indicate lignin (Belitz and Grosch, 1987); volatile sulfur compounds are a marker for sulfated polysaccharides and proteins, each distinguishable by its unique pyrolysis products (Merritt and Robertson, 1967; Sjoberg and Pyysalo, 1985). Carbohydrates alter the pattern of accumulation of aliphatic carbonyl and heterocyclic nitrogen compounds that are suggestive of certain proteins and peptide linkages (Merritt and Angelini, 1971).

X. Volume Fraction

Using the density equation of dispersed polysaccharides at concentrations obeying Raoult's law, ϕ_i may be obtained from V_i vs c_i (Walter and Matias, 1989). When V_i vs c_i is not linear, ϕ_i must be stated at a given c_i. Subsequent to ϕ_i determinations, configurational ΔS [Eq. (3.22)] were calculated for low- and high-methoxyl pectin. Low-methoxyl pectins were discovered to be less inclined to order themselves in an aqueous medium than high-methoxyl pectins (Walter, 1991).

XI. Hydrophilicity

By the same density approach to ϕ_i, the polysaccharide water of hydration was quantified (Walter and Talomie, 1990):

$$\mathbf{H} = [\beta(V_i \cdot \mathbf{d}_i - 1)]/V_i. \qquad (7.13)$$

H, defined as the hydrophilicity, is the weight of hydrocolloidal water in grams adsorbed per gram of solute in 10^2 g of dispersion; β is the slope of the V_i vs c_i graph. Equation (7.13) is based on the proportional nonsolute volume increase when increments of a hydrophilic polysaccharide are added to initially pure water. Not all gums showed the linearity expressed in Eq. (7.13). When there is nonlinearity, e.g., in one sample of konjac flour gum dispersions[18] (Jacon *et al.*, 1993), β is the tangent to the curve at a specific c_i. Of the conforming polysaccharides tested, the decreasing order of hydrophilicity was CMC > guar gum > methylcellulose > sodium alginate > HM pectin > LM pectin. In the opposite direction, dehydration may also not be linear, but exponentially dependent on time (Lips *et al.*, 1988).

XII. Surface Area

In plotting Eq. (2.15) (the Freundlich equation), a series of flasks is called for that contain different weights of adsorbent (grams; e.g., 0.1–1.0 g) in equal volumes of a test dispersion of solute at the same weight concentration. After an equilibrating period of shaking, the supernatant liquid is filtered, its residual concentration c_r is measured, and the filled-sites concentration (amount adsorbed) is obtained by subtraction $(c_i - c_r)$. The graph $(c_i - c_r)/g$ vs c_i at constant temperature gives slope $1/k$. Alternatively, a constant weight of adsorbent (e.g., 1 g) is placed in equal volumes of dispersion at different c_i; $c_i - c_r$ is measured and plotted at each c_i to a plateau region.

Instrumentation is an alternative to chemically measuring surface area. Wheat granule surface area was Coulter-counted (Morrison and Scott, 1986). If a spherical geometry is assumed, Fig. 1 can provide a basis of measuring the total surface area.

XIII. Fiber

The naturally occurring, chemically and physically refractory plant polymers of which cellulose is the dominant component are collectively called fiber. Crude fiber (CF) is the residue remaining after a formal acid and alkali scouring. In an official CF assay (AOAC, 1990), the fibrous material is digested with 1.25% H_2SO_4 followed by 1.25% NaOH, the digest is filtered, the residue is dried, and then it is ignited. The percentage loss in weight is the CF value—the lowest of all possible fiber values, consisting exclusively of cellulose and lignocellulose. CF was once believed to be the analytical equivalent of human and animal indigestible fiber, and a CF assay was therefore the basic analytical criterion of nutritional value, but because it proved to be inadequate for this purpose, new classifications of food and feed fibers succeeded it.

Dietary fiber is defined by nutritionists and physicians as the category of naturally occurring plant components containing soluble and insoluble fiber that increase fecal volume. Soluble fiber consists mainly of pectin and a

18. The manufacturer has since recalled and reengineered konjac flour gum.

fraction of hemicellulose; cellulose, the other gums, lignin, and another fraction of hemicellulose are presumed to be indigestible to humans. Some fiber definitions are based on methodology, e.g., neutral-detergent and acid-detergent fiber.

Fiber analyses require initial fat extraction. The highest fiber value obtainable thereafter is the neutral-detergent fiber (NDF) value, consisting of nothing less than dried plant extract from which emulsifiable lipids and a fraction of protein, pectin, and other carbohydrates have been washed away through the detergent action of sodium lauryl sulfate. In a total dietery fiber (TDF) assay, the lipid-extracted substrate is hydrolyzed by α-amylase (to digest starch), then a protease (to digest protein), then an amyloglucosidase (to hydrolyze branched structures). Insertion of the enzymes in the protocol was an attempt to simulate the action in the human digestive tract (Schaller, 1977). The water-soluble, fibrous fraction (A) is precipitated by alcohol, followed by washing, filtration, and drying of the residue (fraction B). Samples of the dried residue are analyzed for ash by ignition and for protein residue. $A + B$ corrected for residual ash and protein gives the TDF. Replacing the acid, base, and enzyme digestions with refluxing in cetyltrimethylammonium bromide gives acid-detergent fiber (ADF). ADF contains lignin, cellulose, lignocellulose, and insoluble mineral matter, the water-soluble and hemicellulose components having been removed in the supernatant liquid.

XIV. Pilot Plant Quality Control

For practical reasons in a food plant where advanced instrumentation is not always available, it is desirable to have recourse to simple techniques to assess quality. Any property of a polysaccharide dispersion can do this if quality factors are adequately referenced.

A. *Identification*

The most elementary, nontechnical method of identifying a polysaccharide is to burn it, observe the yellow-to-brown color, and sniff the perfumelike aroma that should be reminiscent of the aroma of maple syrup, if the test is positive. If the caramelized residue is shaken with egg white and a visible reaction produces an insoluble, pigmented deposit (melanoidin), it is empirical proof that the sample was a polysaccharide.

Figure 5 Aging of an aqueous, gelatinized starch dispersion showing phase separation with time (10 days).

B. Aging

Freshly prepared dispersions are opalescent and light transmittance, already at a minimum, cannot be improved, nor can the dispersions be clarified by filtration; scattering is at a maximum. The properties are reversed in an aging dispersion, as the particles grow larger and fewer and ultimately separate from the solvent (Fig. 5). Aged dispersions are turbid and filterable.

C. Sediment Volume

Sediment in an aged dispersion may be collected and measured in a crudely quantitative test. To distinguish between a deflocculated, a flocculated, and an aggregated suspension, a weighed amount of solid is uniformly suspended in a small quantity of liquid, the suspension is transferred to a graduate cylinder, the volume of sediment during a stated period of time is measured, and the specific sediment volume (milliliters per gram) vs time is plotted. In such an experiment, a floc's suspension rate was found to be nearly 3 times that of the aggregate's suspension rate and 10 times that of the deflocculated suspension rate (Ross and Morrison, 1988).

D. Syneresis

Alternatively to collecting sediment, liquid may be collected and measured in a conical graduated centrifuge tube from a freshly cut surface of a 200-g sample of jam or jelly; the sample is divided into quadrants on a nylon net. A maximum of 0.5 mL in 2 h is arbitrarily set as an index of good product quality, i.e., a low syneresis tendency (Hercules, Inc., 1985).

E. Consistometry

In Bostwick consistometry, a constant volume of a high-solids suspension (e.g., tomato ketchup, applesauce) is permitted to flow unidirectionally for 30 s along a graduated path in a rectangular, stainless steel trough. Quality grades are preset to correspond with the distance (in centimeters) traversed by the forward edge of the suspension. A US Grade A tomato ketchup requires a distance of 9 cm in 30 s at 20°C.

F. Texture

The most generalized property of a polysaccharide semisolid dispersion is texture, for which there are any number of definitions (Bourne, 1982), each nevertheless suggesting a physiological response to physical stimuli (size, shape, flow, hardness, etc.). Objectively, this elusive property is measured as the force necessary to compress or puncture the test object.

The TA-XT24 texture analyzer (Texture Technologies Corp., Scarsdale, NY) is a device mounted with various mechanical probes, calibrated with weights, and preset to penetrate the test object at variable speeds to variable depths. The magnitude of the force and the shape of the force–time profile (Rao *et al.*, 1989) provide information about fracturability, cohesiveness, springiness, and gumminess of the test material. The Instron Universal Testing machine (Instron Engineering Corporation, Canton, MA) and the Voland texture analyzer (Voland Corporation, Hawthorne, NY) are similar instruments.

XV. Polysaccharide Theta Conditions

Temperature and solvent define the θ state. A polysaccharide lower T_c is in the range of refrigerated temperatures where solute–solvent interaction yields to solute–solute interaction: defining the θ temperature is thus restricted to 25–28°C.

Figure 6 suggests that the viscosity of ionic polysaccharides in dilute *d*-tartaric acid (TA) and of nonionic polysaccharides in water (c_i = 0.05–0.07%) are in the same general $\eta_i - c_i$ orbit at 28°C. A sample of CMC (0.05 g) was dispersed in 80-mL water in 100-mL beakers to which TA was afterward added to different molarities; TA supplied the H^+ counterion intrinsic to an ionic polysaccharide and the nonintrusive tartrate ion. The solutions were transferred to 100-mL Erlenmeyer flasks and brought to volume with water prior to dilution viscometry. Judging from Fig. 7, a molar concentration of TA approximating 0.35 ensures an η_{sp}/c_i minimum in a dilute CMC dispersion (c_i = 0.05–0.07%).

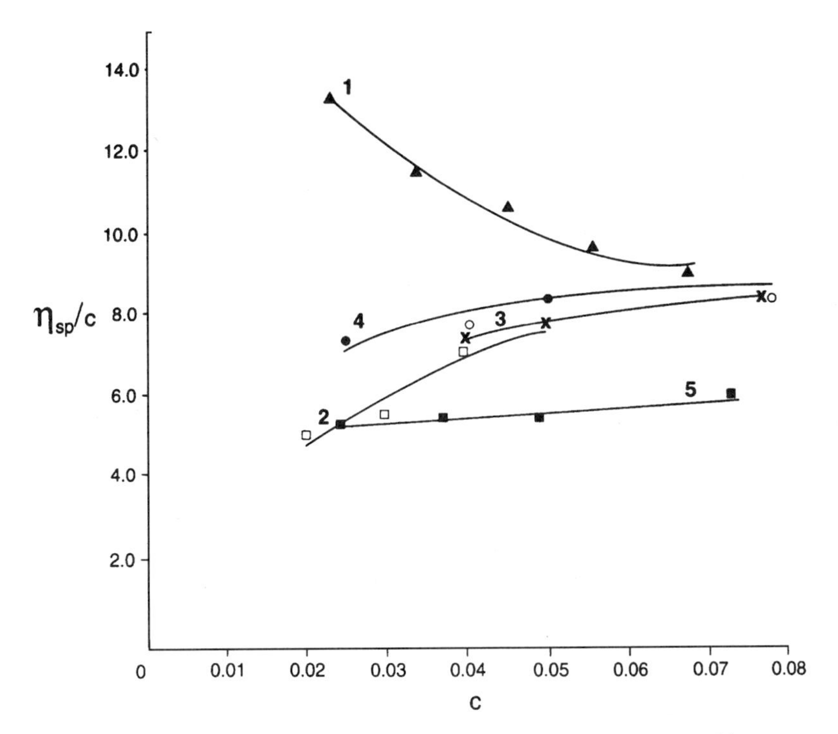

Figure 6 Viscosity profile of gellan in water (1), gellan in 0.04-M tartaric acid (2), locust bean (3) and methylcellulose (4) in water, and CMC in 0.04-M tartaric acid (5).

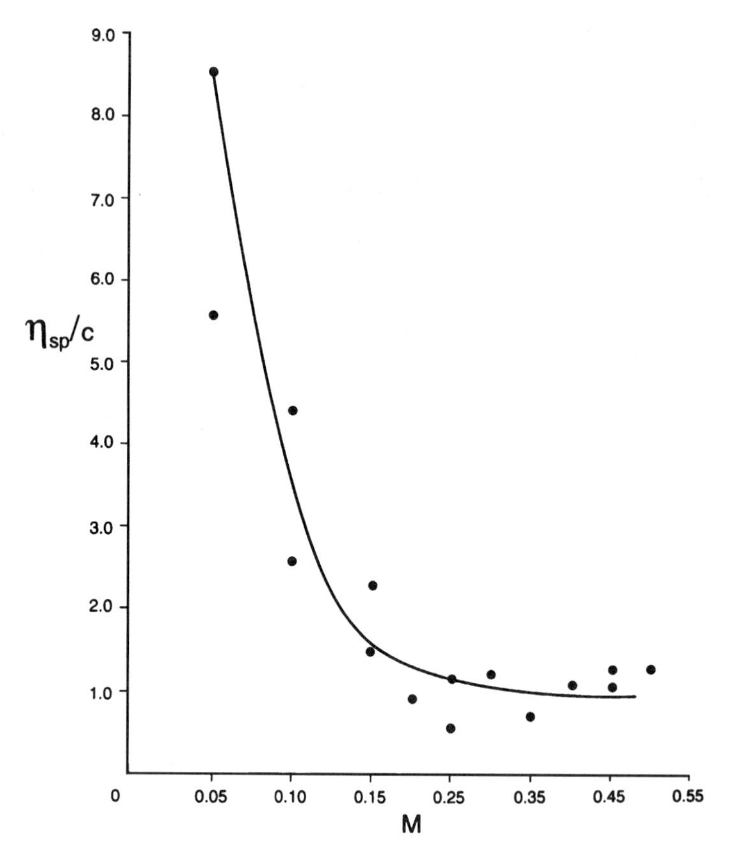

Figure 7 Viscosity profile at 28°C of 0.05% CMC in different molal concentrations (M) of tartaric acid.

From Alfrey's (1947) analysis, the region of the negative slope of polyanions before the minimum viscosity function (gellan in Fig. 6 and CMC in Fig. 7) is a region of declining solute–solvent interaction, followed by an increasing solute–solute affinity; the latter continues through to a positive slope after the minimum.

On the strength of the results in Fig. 7, a second series of dispersions was similarly made with different concentrations of CMC in 0.35-M tartaric acid (Fig. 8). Flow times were 57 s for 0.05% CMC in 0.35-M tartaric acid and 76–96 s for the range of CMC dilutions. There was no minimum in evidence at dilute CMC concentrations, and as seen in Fig. 8, η_{sp}/c_i increased directly albeit slowly with c_i, which is presumptive evidence that solute–solute

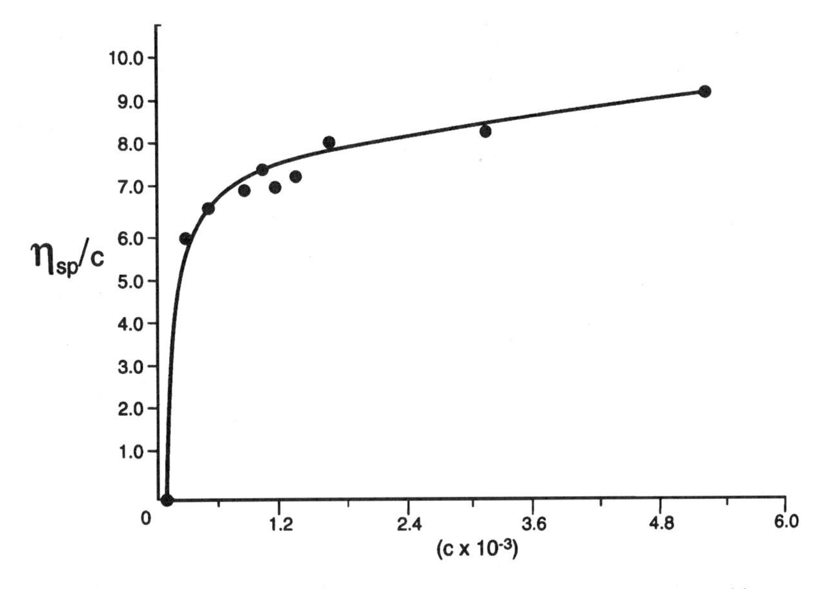

Figure 8 Viscosity profile at 28°C of aqueous CMC at different concentrations (*c*) in 0.35 M tartaric acid solution.

interaction exists in dilute (CMC) dispersions. However, the viscosity number decrease [$\eta_{sp}/c = (9.5–7.7)$] was merely 20% for a quadrupling of the concentration ($c = 4.8–1.2 \times 10^{-3}$). Given interaction on either side of the minimum, the θ state is arguably unachievable with polysaccharides, but may be approximated. The most that can be said of these data is that the θ condition is approached at concentrations in the third decimal. Neutral polysaccharides displayed similar relationships, omitting electroviscosity.

The effect of 40°C on CMC dispersions was studied and found to be quite similar to the effect at 28°C (Fig. 9). The 40°C line gave a correlation coefficient of 0.81. When the point at 4.0×10^{-4} was removed, the correlation coefficients rose to 0.95, suggesting marginal sensitivity at that level of dilution. Figure 10 shows that the effect of a hydrophobic solution environment was identical on a dilute ionic (CMC) and a neutral polysaccharide at the same dilute concentration. In each instance, the slope of the line approximated 0. Recognizing the dilution effect on polyanion viscosity, it may be argued that ethanol is as efficacious as electrolytes in inducing not only nonionic behavior in dilute polysaccharide polyanions, but in controlling solute–solvent interaction in preparation for $\overline{M}$ and R_g measurements.

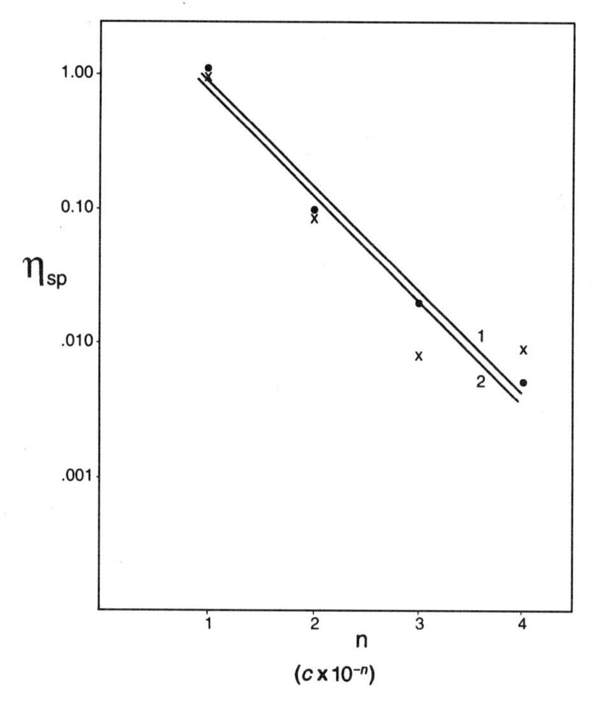

Figure 9 The effect of temperature (28°C, line 1 and dots, and 40°C, line 2 and multiplication signs) on the specific viscosity (η_{sp}) of a 0.05-wt/vol% CMC dispersion.

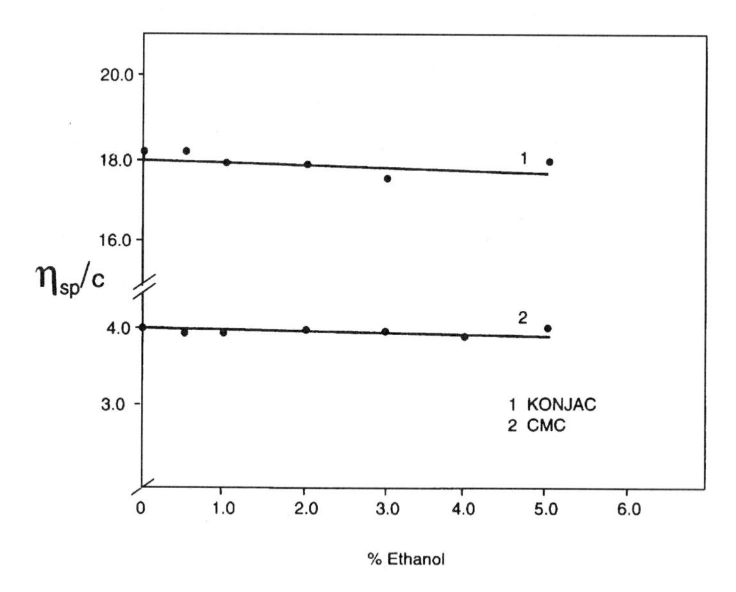

Figure 10 The effect of ethanol (vol/vol%) on the viscosity number (η_{sp}/c) of a 0.05-wt/vol% konjac (1) and CMC (2) dispersion.

XVI. Blending

The blending chart in Fig. 1 in Chapter 5 shows an exponential dependence of solution viscosity on concentration, within the range of maximum measurable concentration (100%). Rheometry permits higher c_i than does viscometry. As a prerequisite to construction of a linear polysaccharide blending chart, the linear range of η_{sp}/c_i vs c_i should first be established, because it is only then that the viscosity of the blend will be equal to the sum of the viscosities of the components. Each ordinate in Fig. 1 in Chapter 5 is labeled with the maximum c_i of the two polysaccharides of interest or its equivalent as 100%.

XVII. Summary

Many properties in common often make polysaccharides analytically indistinguishable from each other, except for differences occasioned by ionic and nonionic composition. However, subtle differences occasionally exist, so that a specific test may be designed for a specific molecule.

Polysaccharides almost always require tedious isolation and purification techniques to insure a high degree of relative purity before characterizing methodologies are applied. The uniqueness of the hydrocolloidal response to ambient stimuli enables them to be characterized nondestructively in their dispersion surroundings. Concentration dependence, solvent conditions, thermal effects, transitions, etc., are defining variables of their functional utility. Reactive chemicals, heat, and critical time–temperature integrals partially or totally decompose them. The decomposition products are useful as flavorants, colorants, and chemical markers. Especially in the presence of nitrogen compounds, pyrolysis yields the universally pleasant aromas of bread, cakes, and nuts. Carbohydrate and carbohydrate–protein thermolytic reactions are the cause of the yellow-to-brown coloration observed in many cooked carbohydrate foods.

The size of a polysaccharide molecule is an indeterminate property: molecular weight, for example, depends on the analytical method. Electroviscosity and solute–solvent and solute–solute interactions affect the volume–gram relationship (density) and conformational dimensions. The degree of polymerization is not affected and is therefore a constant, reliable indicator of the molar mass.

Elementary quality control tests involving a visible polysaccharide property or response can suffice in quality control laboratories for the lack of advanced instrumentation.

CHAPTER **8**

Classifications

I. Introduction

Polysaccharides are broadly divisible into two main groups—ionic and nonionic. Otherwise they possess many properties in common that transcend empirical class boundaries. They are systematically classified according to origin, isolation method, function, texture, thermoreversibility, and gelling time (Trudso, 1989) based on the classifiers' interest: for example, the science nomenclator organizes polysaccharides by chemistry, but to the food technologist, a class of nongelling polysaccharides suggests that an aqueous dispersion of any member of the class does not solidify under food processing and preparation conditions, where functionality is the primary interest. By polysaccharide functionality is meant the ability of a polysaccharide to impart properties to or assist in improving edible matter from the standpoint of human nutritional and food-processing quality. Tables I and II are an indication of the uses and applicable concentration ranges of common food polysaccharides.

II. Chemical Classification

Chemical classification of polysaccharides is the least ambiguous system of grouping these macromolecules. Polysaccharides of different origins can have similar structures, as Kravtchenko *et al.* (1992) discovered in lemon and apple pectin. Likewise, polysaccharides of the same origin can have different structures, as for example, the concentration of pyruvate and acetyl substituents in xanthan, depending on incubation conditions (Pettitt, 1982).

TABLE I
Properties and Uses of Polysaccharides in Food[a]

Property	High methoxyl pectin	Low methoxyl pectin	Kappa carrageenan	Iota carrageenan	Lambda carrageenan
Solubility in water	Sol. cold and hot	Sol. cold and hot	Sol. above 70°C, Na^+ and NH_4^+ Sol. cold	Sol above 70°C. Na^+ and NH_4^+. Sol. cold. K^+ and Ca^{2+} swell cold to thixotropic dispersions	Sol. cold and hot
Solubility in milk	Sol. cold and hot	Sol. hot	Sol. above 70°C	Sol. above 70°C	Sol. hot. Swells cold
Solubility in salt solutions	Insoluble	Insoluble	Insoluble	Sol. hot	Sol. hot
Solubility in sugar solutions	Sol. hot	Sol. hot	Sol. hot	Insoluble	Sol. hot
Solubility in ethanol	Insol. above 20%	Insol. above 20%	Insol. above 20%	Insol. above 20%	Insol. above 20%
Other factors influencing solubility	Increases with decreasing MW, increasing randomness of COOH, decreasing sugar and Ca^{2+}	Increases with decreasing sugar and Ca^{2+}	Increases with decreasing Na^+, K^+ and, Ca^2	Increases with decreasing Na^+, K^+, and Ca^{2+}	Increases with decreasing Na^+, K^+, and Ca^{2+}
Solution viscosity	Low	Low	Low	Medium	High
Optimum pH range	2.5–4.0 pK_a 3.3	2.5–5.5	4–10	4–10	4–10
Optimum soluble solids range	55–80%	30–80%	0–40%	0–20%	0–80%

Gelation conditions	pH below 4 and sol. solids 55–80%	Presence of Ca^{2+} 10–70 mg/g pectin. Temp. below setting temp.	Presence of K^+, Na^+ or Ca^{2+}. Temp. below setting temp.	Presence of K^+, Na^+ or Ca^{2+}. Temp. below setting temp.	Nongelling
Gel characteristics:					
Texture	Cohesive no syneresis, thermoirreversible	Cohesive to brittle. Brittleness increases with increasing Ca^{2+} and decreasing sugar. Thermoreversible	Strong, brittle. Brittleness increases with increasing K^+ and Ca^{2+} and decreasing LBG. Thermoreversible	Soft, cohesive, thixotropic. Thermoreversible Thixotropy is lost with addition of LBG	Nongelling
Setting temp.	Increases with increasing DE, decreasing pH and increasing sugar	Increases with decreasing DE, increasing DA, increasing Ca^{2+} and increasing sugar	Increases with increasing K^+, Na^+, Ca^{2+}, and sugar	Increases with increasing K^+, Na^+, Ca^{2+}, sugar, and LBG	Nongelling
Gel strength	Increases with increasing concentration and MW	Increases with increasing concentration and Ca^{2+}	Increases with increasing concentration, K^+, Ca^{2+}, and LBG	Increases with increasing concentration, K^+, Na^+, and Ca^{2+}	Nongelling
Effect on milk at neutral pH	Precipitation	Gelation	Ionic interaction. Increased gel strength	Ionic interaction. Increased gel strength	Ionic interaction. Increased viscosity
Effect on milk and other proteins at acid pH	Adsorption to casein particles below pH 4.2. Adsorption to soy protein particles below pH 4.8.	None	Precipitation below iso-pH	Precipitation below iso-pH	Precipitation below iso-pH

continued

TABLE I
continued

Property	High methoxyl pectin	Low methoxyl pectin	Kappa carrageenan	Iota carrageenan	Lambda carrageenan
Incompatibility	Water soluble alcohols, ketones, heavy metals, quaternary detergents, cationic macromolecules	Water soluble alcohols, ketones	Water soluble alcohols, ketones, quaternary detergents, cationic macromolecules	Water soluble alcohols, ketones, quaternary detergents, cationic macromolecules	Water soluble alcohols, ketones, quaternary detergents, cationic macromolecules

Property	Guar gum	Xanthan gum	Gelatin	Gum arabic
Solubility in water	Sol. cold and hot	Sol. cold and hot	Sol. above 40°C	Sol. cold and hot
Solubility in milk	Sol. cold and hot	Sol. cold and hot	Sol. above 40°C	Sol. cold and hot
Solubility in salt solutions	Sol. cold and hot	Sol. cold and hot	Sol. above 40°C	Sol. cold and hot
Solubility in sugar solutions	Sol. cold and hot	Sol. cold and hot	Sol. above 40°C	Sol. hot
Solubility in ethanol	Insol. above 20%	Insol. above 50%	Insol. above 20%	Insol. above 60%
Other factors influencing solubility			Increases with decreasing MW	Increases with increasing pH up to 6
Solution viscosity	High cold. Low hot	High below 100°C	Low	Low
Optimum pH range	4–10	1–13	4.5–10 iso-pH 4.8–5.2 (limed) iso-pH 6.0–9.5 (acid)	2–10

Optimum soluble solids range	0–80%	0.80%	0.80%	0.80%
Gelation conditions	Nongelling	Presence of LBG, tara gum, cassia gum. Temp. below setting temp.	Temp. below setting temp.	Nongelling
Gel characteristics:				
Texture	Nongelling	Cohesive, gummy, thermoreversible. Guar makes texture of xanthan/LBG gel more brittle	Soft to strong, cohesive, gummy. Thermoreversible	Nongelling
Setting temp.	Nongelling	Constant	Increases with increasing MW and maturing temperature	Nongelling
Gel strength	Nongelling	Increasing with increasing concentration	Increases with increasing concentration and decreasing salt	Nongelling
Effect on milk at neutral pH	Separation	None	None	None
Effect on milk and other proteins at acid pH	None	Precipitation below iso-pH	None	None
Incompatibility	Water soluble alcohols, ketones	Water soluble alcohols, ketones, gum arabic below pH 5	Water soluble alcohols, ketones, anionic macromolecules below iso-pH, gum arabic below iso-pH	Water soluble alcohols, ketones, alg., gelatin, xanthan gum

continued

TABLE I
continued

Property	Agar-agar	Alginate	Propylene glycol alg.	Cellulose gum	Locust bean gum
Solubility in water	Sol. above 90°C	K^+, Na^+, NH_4^+ Sol. cold and hot Ca^{2+} Insol. at neutral pH	Sol. cold and hot	Sol. cold and hot	Sol. above 85°C
Solubility in milk	Sol. above 90°C	Insol. Na^+ swells in boiling milk. Sol. with sequestering agents	Sol. cold and hot	Insoluble	Sol. above 85°C
Solubility in salt solutions	Sol. above 90°C	Insoluble	Insoluble	Insol. High DS types sol.	Sol. above 85°C
Solubility in sugar solutions	Sol. above 90°C	Sol. hot	Sol. cold and hot	Sol. cold and hot	Sol. above 85°C
Solubility in ethanol	Insol. above 20%	Insol. above 40%	Insol. above 40%	Insol. above 30%	Insol. above 20%
Other factors influencing solubility		Increases with decreasing COOH, increasing pH, decreasing Ca^{2+}		Increases with decreasing MW, increasing pH, increasing divalent cations	
Solution viscosity	Low	Low above pH 5.5 High below pH 5.5	High	High	High up to 85°C
Optimum pH range	2.5–10	2.8–10, pK_a 3.4–4.4	2.8–10	3–10 pK_a 4.2–4.4	4–10

Optimum soluble solids range	0–80%	0–80%	0–80%	0–80%	0–80%
Gelation conditions	Temperature below 32–39°C	pH below 4 or presence of Ca^{2+} 20–70 mg/g alg.	Nongelling	Nongelling (gelation may occur with trivalent cations)	Nongelling
Gel characteristics:					
Texture	Strong, brittle. Thermoreversible. Brittleness increases with increasing sugar	Acid gels soft, cohesive and thixotropic. Calcium gels strong, brittle, Thermo-irreversible	Nongelling	Nongelling	Nongelling
Setting temp.	Constant	Nonexistent	Nongelling	Nongelling	Nongelling
Gel strength	Increases with increasing concentration, increasing sugar and increasing pH	Increases with increasing concentration, Ca^{2+} and decreasing pH down to 3.6	Nongelling	Nongelling	Nongelling
Effect on milk at neutral pH	None	None. Insoluble	None	Precipitation	Separation
Effect on milk and other proteins at acid pH	None	None	None	Adsorption to casein particles below pH 4.6. Adsorption to soy protein particles below pH 5.0	None
Incompatibility	Water soluble alcohols, ketones	Water soluble alcohols, ketones, milk, gum arabic	Water soluble alcohols, ketones	Water soluble alcohols, ketones, quaternary detergents, cationic macromolecules	Water soluble alcohols, ketones

[a] From Trudso, 1989. Reprinted with permission.

TABLE II
Function and Concentration Ranges of Polysaccarides in Food.

Hydrocolloid	Food	Function	Concentration (%)
Pectin	Jams, jellies, preserves	Gelation, thickening	0.1–1.0
	Bakery fillings, glazings	Gelation, thickening	0.5–1.5
	Fruit preparations	Thickening, stabilization	0.1–1.0
	Fruit beverages, sauces	Thickening, stabilization	0.01–0.5
	Confectionery	Gelation, thickening	0.5–2.5
	Dairy products	Stabilization, gelation	0.1–1.0
Carrageenan	Ice cream	Stabilization	0.01–0.03
	Chocolate milk	Stabilization	0.01–0.03
	Flans and puddings	Gelation, thickening	0.1–0.5
	Liquid coffee whitener	Thickening	0.1–0.2
	Low calorie jams	Gelation, thickening	0.8–1.2
	Dessert gels	Gelation	0.6–1.1
	Tart glazing	Gelation	0.8–1.0
	Meat products	Gelation, water binding	0.3–0.5
	Pimiento paste	Gelation	1.5–3.0
	Salad dressing	Stabilization	0.3–0.6
Agar-agar	Icings	Gelation	0.1–0.3
	Confectionery	Gelation	0.3–1.8
	Meat products	Gelation	0.5–2.0
	Dairy products	Gelation	0.05–0.9
Alginate	Ice cream	Stabilization	0.1–0.5
	Icings	Gelation	0.1–0.5
	Toppings	Gelation	0.3–0.5
	Salad dressings	Stabilization	0.2–0.5
	Beer	Stabilization	0.004–0.008
	Fruit drinks	Stabilization	0.1–0.3
	Restructured foods	Gelation	0.6–1.0
	Simulated fruit	Gelation	0.8–1.0
CMC	Ice cream	Stabilization, thickening	0.1–0.3
	Ripples	Thickening	0.1–0.4
	Sour milk	Stabilization	0.1–0.2
	Cake mixes	Moisture retention	0.2–0.4
	Icings	Thickening, water binding	0.1–0.2
	Batters	Thickening, stabilization	0.2–0.4
	Dry-mix beverages	Thickening	0.1–0.3
	Syrups	Thickening	0.2–0.6
LBG	Ice cream	Stabilization	0.2–0.3
	Cream chesse	Thickening, moisture control	0.3–0.6
	Dessert gels	Gelation, water retention (together with carrageenan)	0.3–0.6
Guar gum	Ice cream	Stabilization	0.2–0.3
	Cottage cheese	Thickening	0.3–0.6
	Processed cheese	Moisture retention	0.2–0.4
	Cake mixes	Thickening	0.1–0.2
Xanthan gum	Bakery jellies	Gelation, thickening	0.1–0.3
	Fruit drinks	Pulp suspension	0.02–0.06
	Cream cheese	Gelation	0.1–0.2
	Baked goods	Moisture retention	0.1–0.2
	Dressings	Stabilization	0.2–0.3

TABLE II
continued

Hydrocolloid	Food	Function	Concentration (%)
Gelatin	Yogurt	Gelation	0.3–1.0
	Dessert gels	Gelation	4–6
	Confectionery	Gelation	3–10
Gelatin	Meat products	Gelation	1–5
	Mousses	Stabilization	1–3
	Minarine	Stabilization	1–3
Gum arabic	Flavor fixation	Encapsulation	80–90
	Confectionery	Stabilization, gelation	10–60
	Flavor emulsions	Stabilization, emulsification	10–30

[a]From Trudso, 1989. Reprinted with permission.

A. α-D-*Glucans*

This group of nonreducing glucose polysaccharides contains a preponderance of 1,4-α-D-glycopyranosyl linkages. The most ubiquitous representative is starch.

1. Starch

Native starch (Whistler and Smart, 1953; Biliarderis, 1992) is deposited in granules in plant tissues and is constructed of amylose (1,4-α-D bonding) and amylopectin (1,4-α-D bonding in the primary structure and 1,6-α-D bonding at branch points). Potatoes and legumes are among the highest natural accumulators—34–70% in peas (Zuber, 1965), 20–36% in ordinary corn, and 16–17% in rice. Granule sizes and morphologies are typical of the genus, from irregular 10–25-μm-diameter polyhedrons in corn to uniform polyhedrons in rice (3–8 μm), and spheroids in potatoes (15–100 μm; 33 μm, average). Other geometries encountered are ellipsoids, polygons, platelets, and tubules (Roller, 1996). The proportion of amylose to amylopectin varies with the plant source; normally it is about 1:3. Waxy maize and waxy rice contain little or no amylose: waxy maize starch cooks to a clear, stable, nongelling paste, whereas regular corn starch cooks to a cloudy, retrograding gel. Amylomaize is starch from corn that has been bred to contain as much as 70% amylose. The optical anisotropy resulting from morphology creates unique polarization crosses that are generally distinguishable by phase microscopy. Amylose was cited as an example of a non-free-draining random coil (Ring and Whittam, 1991).

Starch is considered to be a neutral molecule, but potato starch has a low degree of phosphation and is thereby endowed with weak polyelectrolytic character. Starch phosphate esters reside principally in the amylopectin fraction.

The mechanical integrity of starch granules is inversely related to size: small granules are more resistant to dry heat than are large granules. During extrusion, corn and potato starch, containing the largest granules, disintegrate more than did rice starch (Mayer, 1993).

Given the numerous plant origins and macromolecular heterogeneities (composition, granule size, shape, etc.), starch gelatinizes, not at a single temperature, but over a narrow range of temperatures, commonly 50–80°C, whose midpoint is cited as T_{gz}. Each generic starch has a characteristic T_{gz}. Concentration, the amylose–amylopectin ratio, the DP, pH, electrolytes, and cosolutes (sugars) are other influences on the exact location of T_{gz}. The smaller the DP, the lower is T_{gz}; this is explained by the greater kinetic activity of smaller particles. Starch granules remain refractory assemblies, once the temperature is maintained below T_{gz}. Suspended wheat granules held at T_{gz} for 72 h increased in size, porosity, and adsorbent properties while retaining their anisotropy; subsequently they gave a narrower T_{gz} range and higher T_{gz} than the control granules (Gough and Pybus, 1971). Of the common starches, high-amylose corn and rice starch have the highest T_{gz} (74.5 and 80°C, respectively); barley and tapioca have the lowest (57°C); wheat (61°C) and potato (63°C) are intermediate (Glicksman, 1969). For the same intrinsic reason of a variable T_{gz}, starch sols show widely differing gelling concentrations in water. Potato starch, typically containing the largest granules, gels at 20% concentration in water; waxy corn starch gels at approximately 30% concentration (Osman, 1967); corn starch gels at approximately 5% concentration.

Gelatinization is reversible in the first stage (hydration). As the event progresses, there is a gradual loss of polarization and an irreversible loss of birefringence. Gelatinized starch is more susceptible to acid, heat, and enzymes than granular starch. Cooking makes starchy foods digestible as a result of gelatinization.

There is some debate about the role of amylose in crystallite formation, because of its great mobility (Rutenberg, 1980). Contrary to the expected correlation between parallelism and crystallinity, Banks and Greenwood (1975) and Cairns *et al.* (1991) found evidence to implicate principally amylopectin, rather than amylose. Crystals did remain after amylose was leached from the granules (Montgomery and Senti, 1958). Although waxy starches are highly branched, they show signs of localized organization where there is a high concentration of closely packed, linear segments.

The rate of retrogradation is time-dependent and inversely temperature-dependent. The opacity developed in stored starch sols, the staling of bread, etc., are retrogradation phenomena initiated by aging, progressing at a faster rate at a lower temperature. Deterioration is repressed by sodium chloride (Rutenberg, 1980) and sugar (Miles *et al.*, 1985), and is customarily delayed in starch dry mixes by diluents (sugar), anticaking additives (calcium citrate, aluminum silicate, acid phosphate), and crystallization inhibitors (ungelatinized starch, protein, hemicellulose, invert sugar, and other hydrocolloids). Retrograded starch will partly or wholly redisperse by adequate heating in a moist atmosphere; the softening of bread by microwave reheating is based on this principle.

An amylogram of granulated starch shows viscosity passing through a maximum (the peak viscosity) at T_{gz} and falling precipitously, if the temperature is raised further; this response is also exhibited during shearing. The collapse of the granule and tertiary structures, a result of too much mixing, explains the concave upper surface seen in failed breads and cakes. A superior quality of breads, cakes, and pastries contains a fair percentage of intact granules that helps to strengthen the batter or dough against thermal collapse. In a subsequently cooled starch paste, a weak structure is reformed and a secondary viscosity maximum called setback is observed. The higher the amylose content of starch, the higher is the setback on the numerical scale of the amylograph.

In the manufacture of "instant" food items, starch is precooked in a moist atmosphere, then dehydrated and formulated with compatible ingredients into dry mixes. At serving time, water or milk is added, with or without mild heating, and the instantly rehydrated product is ready to be consumed. The broken-curve heating profile of intermediate-moisture, thixotropic starchy products pictorializes gelatinization and solid-liquid (gel–sol) transitions. At high starch concentrations, the initial suspension heats by convection, because of the high heat conductivity of water. Subsequently it heats by conduction, as the gelatinized starch is released into the aqueous medium, thickening it to a gellike consistency. An initially solid mass may revert to a liquid mass; then heat convection and conduction are reversed.

Qu and Wang (1991b) ascribed sensorial appeal to the "melting" of starch granules and the ratio of gelatinized to melted starch. Other factors to consider are pH, oral stress, temperature, and concentration (Olkku, 1978).

For mechanical stability against hydration and swelling, starch is chemically crosslinked (by adipic acid–acetic anhydride, phosphate, etc.) below T_{gz}, so that the physical integrity of the granules is not impaired by gelatinization. Weakly crosslinked starches ($DS < 0.1$) are acid-stable and thus find occasional application in recipes containing vinegar. The many industrial functions of crosslinked starch include use as an adhesive.

2. *Modified Starches*

Substituent groups have been incorporated into dry or suspended starch, mainly to improve its properties and to stabilize it against retrogradation: dispersion clarity is enhanced by substitution (Wurzburg, 1995).

Acetylated starch is more mechanically stable, has a lower cooking (swelling and gelatinization) temperature, and is less prone to gel than the parent starch. Phosphated starch has the advantages of acetylated starch and the extra advantage, in some instances, of high viscosity. The carboxylated, sulfated, and phosphated starches suffer from electrolyte sensitivity. Hydroxyethylstarch and hydroxypropylstarch are freeze–thaw, acid-stable ethers more indifferent to long heating times and high temperatures than other derivatized food starches. Cationic starches are the product of chemical derivatizations with nitrogen-, sulfur-, and phosphorus-containing reactants; they become positively charged and therefore useful as sequestrants of organic and inorganic anions. Many other substituted starches have applicability in food and nonfood industries (Wurzburg, 1986). Actual and potential uses include sizing textiles and paper and as adhesives soluble in organic solvent.

3. Dextran

Dextran is a structurally heterogeneous glucan containing linear and branched sequences with a preponderance of 1,6-α-D (isomaltose) and a smaller percentage of 1,2-α, 1,3-α, and 1,4-α-D linkages. This heterogeneity results in easy dispersion in hot and cold water. Insolubility is associated with 1,3-α linkages (Sidebotham, 1974).

The most important application of dextran has been in medicine, where the 10^4–10^5-Da homologs are used as a blood-plasma substitute. The main industrial use is as a crosslinked gel bed, sulfated for use in ion exchange. Dextran has emulsion-stabilizing properties at low ionic strength, and creaming properties at high ionic strength (Dickinson *et al.*, 1989); it can be a problem in sugar factories where spontaneous fermentation can plug filters and hinder crystallization.

4. Pullulan

Pullulan is a linear, water-soluble polysaccharide containing repeating trimers of 1,4-α-D-glucopyranose and 1,6-α-D-glucopyranose in a 2:1 ratio. Its films are touted for their similarity to polyethylene (Glicksman, 1982).

B. β-D-Glycans

Cellulose (Ott, 1943; Corbett, 1963; Green, 1963) is the most dominant biopolymer in this class.

1. Cellulose

Cellulose is the 1,4-β-D anomer of starch. It has a repeating cellobiose dimer that exists *in vivo* as closely packed 2–30-nm-long microfibrils embedded in a noncellulose matrix. The microfibrils are thoroughly hydrated in succulent fruits and vegetables in which moisture can be as high as 95%.

Once dehydrated, the microfibrils are practically without functionality in ordinary food processing and preparation operations, because the inert microcrystallites are difficult for water to penetrate. The polymorphs, cellulose I and II (Blackwell, 1982; Coffey *et al.*, 1995), are differentiated by their molecular orientation, hydrogen-bonding patterns, and unit-cell structure. Cellulose I is the natural orientation; cellulose II results from NaOH treatment under tension of cellulose I with 18–45% alkali (mercerization). The I–II transition is irreversible. Mercerization strengthens the fibers and improves their lustre and affinity for dyes (Sisson, 1943). Sewing thread was relatively pure mercerized cotton until the advent of synthetic polymer fibers.

Cellulose is designated α, β, and γ on the basis of alkali solubility. α-Cellulose is that fraction not removed by treatment with 17.5% NaOH at 20°C; the 17.5% NaOH-soluble fraction contains β- and γ-cellulose. The subfraction precipitating after acidification of the alkaline liquid phase is β-cellulose; γ-cellulose is the acid- and base-soluble subfraction remaining dispersed. α-Cellulose contains the highest DP.

Native cellulose is engineered into a variety of utilitarian forms (Fig. 1). Cotton is the cellulose fibers harvested from around the seeds of *Gossypium sp.*: it experiences an infinite number of applications, chiefly in the textiles and medical industries. Solka-floc is purified, shredded, fluffed wood cellulose noted for its absorbency. Paper is the final product of wood cellulose that is mechanically disintegrated to free the fibers, agitated to intertwine them and strengthen the fiber network, and pressed to smooth the network surface. Mechanical pulp may be chemically treated further to remove impurities before formulation to desired specifications with additives. By steeping shredded-wood alkali-cellulose in CS_2 (xanthation) and extruding the viscous fluid (viscose) three dimensionally as fibrils (rayon) or two dimensionally as sheets (cellophane) into an acid bath, a translucent cellulose gel is regenerated. This regenerated cellulose, per se, when dry, is both grease- and oil-proof and, for all practical purposes, gas impermeable (Flexel, 1989), but it does not provide moisture protection. Plasticizers, sealers, sizers, and other chemicals are added to viscose to confer miscellaneous properties on cellophane. For some uses, nitrocellulose and polyvinylidene chloride are layered onto the surface of the sheets to improve their poor moisture barrier property (Reiter, 1986). Figure 1d is cellophane coated with saran (a vinylidine copolymer) for moisture and chemical resistance and thermoplasticity (heat sealability). An essential difference between cellophane and pulp is the absence of crystalline sites in the cellophane; they are removed by alkalization and xanthation, which consequently intensifies the transmitted light vis-à-vis the scattered light. A certain high quality of paper (bond) scatters light profusely, due to the inclusion of titanium dioxide as a coating and filler additive.

Cellulon is the trade name of a cellulose newly developed from bacterial fermentation. It is reported to have unique properties as a result of its small

Figure 1 Utilitarian forms of cellulose.

fiber size (0.1 μm diameter), which enables it to expose 200 times more surface area than other cellulose fibers. Cellulon is claimed to have great binding, shear-thickening, and coating power at very low concentrations, and to be synergistic with other thickeners (Krieger, 1990).

The amorphous regions of cellulose are less chemically and physically reactive than their starch counterparts, but may nevertheless be hydrolyzed to a "level-off" or limiting concentration of 40–70% of intact, pure, rodlike crystallites several hundred angstroms long and less than 100 Å wide (Tanford, 1961). Limit cellulose facilitates the prerequisite rodlike ordering for optical anisotropy and precipitation. Limit cellulose is further partially degraded by mechanical attrition to a microcrystalline DP of 200–300 (Battista and Smith, 1962), sold under the brand name Avicel (FMC, 1993). Avicel in aqueous suspension exhibits weak hydrophilic character and, by stiffening films and foams, it is a strong transparent barrier to lipid migration. It stabilizes foams by organizing itself in tandem into gellike, fibrillar, heat-stable structures around bubbles of air (FMC, 1993). Thermophysical and thermochemical refractoriness make Avicel a premier foam stabilizer of retorted polysaccharide foodstuffs.

The life of an Avicel suspension can be extended by coprecipitating the rodlike structures with a protective colloid after trituration. Avicel-RC[19] is limit cellulose that has been physically modified by coprecipitation with CMC to facilite dispersibility. Avicel-RC water suspensions simulate the properties of a hydrosol. At low aqueous concentrations, the apparently hydrated crystallites assemble into a thixotropic, heat- and acid-stable structure whose viscosity depends directly on pH to about pH 10, whereupon it declines precipitously. The suspension coalesces at low pH. The addition of salt after mixing increases viscosity above what it would be if the salt were added at the time of mixing or shearing.

19. RC is an abbreviation for regenerated cellulose.

A microdefibrillated suspension of cellulose in water has been reported to have an indefinite shelf-life (Weibel, 1994).

2. *Cellulose Derivatives*

Controlled derivatization to various DS as low as 0.05 causes cellulose to lose its crystalline habit. The family of these derivatives is known collectively as cellulose gums. The properties of cellulose gums are a function of the nature, quantity, and distribution of the substituent(s); their rheology is affected much less by ester and ether content than by molecular weight (Krumel and Sarkar, 1975). Most cellulose derivatives are acid-stable over the normal range of food acidities and, unlike other polysaccharides, a number of them gel when heated, and return to the sol state when cooled: the gelation mechanism is crystallization (Sarkar, 1979). Alkyl and hydroxyalkyl substituents cause cellulose derivatives to tolerate high concentrations of ethanol (20%). Given the inherent stiffness of the cellulose primary structure, these derivatives tend to behave as liquid crystals (Stannett, 1989), i.e., they are ordered at rest and disordered when disturbed.

Methylcellulose (Dow Chemical Co., 1990) is a water-soluble, thermostable ether that gels reversibly upon heating at 40–50°C. Sucrose lowers T_{gel}. The gels are prone to hysteresis (Grover, 1993). This biopolymer is an excellent foam stabilizer, as demonstrated in Fig. 1 in Chapter 2 by the visible retention of air bubbles at the solid–liquid interface.

Hydroxypropylcellulose is also a water-soluble ether that is acid-stable, strongly lipophilic, and has gelling characteristics identical to those of methylcellulose. Gels from the lower DP homologs are harder, more rigid, and more dimensionally stable than gels from higher DP homologs (Desmarais and Wint, 1993).

Food-grade CMC is a cellulose carboxylic acid ether with an optimum DS = 0.4–0.7. The higher the DS within this range, the more hydrophilic is the polyanion. Uniformity of substitution makes CMC more compatible with dissolved salts and less inclined to thixotropy than uneven distribution (Feddersen and Thorp, 1993). This gum does not precipitate from a 50% ethanol solution. Below approximately pH 4 in water, the polyanions revert to the un-ionized, water-insoluble acid. CMC viscosity–hysteresis has already been described (Fig. 2 in Chapter 3). CMC dispersions and films have the extra advantage of transparency relative to many other polysaccharide dispersions. The films are resistant to oils, grease, and organic solvents (Hercules, Inc., 1980).

Chitin is the naturally occurring, insoluble acetylamino derivative of cellulose yielding the polycation chitosan when deacetylated (with concentrated NaOH and heat). Chitosan is water-soluble in proportion to the degree of deacetylation.

3. Curdlan

Curdlan (Kimura *et al.*, 1973; Sandford, 1979; Nakao *et al.*, 1991; Harada *et al.*, 1993) is a 1,3-β-D-glucan whose cloudy aqueous suspensions develop a soft, thermosetting, freeze–thaw-stable, translucent gel when heated above 80°C. Gelation may be induced by the action of Ca^{2+} in weakly alkaline dispersions. Curdlan gels are hard and brittle in the acidic pH range, but soft and elastic in the alkaline pH range. Gel strength increases with increasing temperature and duration of heating. Curdlan gel properties are exploited in the fabrication of a wide assortment of textured foods, with the proviso that they be protected from premature exposure to heat, because of the flow irreversibility of the thermoset. Curdlan xerogels swell in cold water but do not disintegrate.

4. Glycomannans

This group of polysaccharides is represented by locust bean and guar gums (1,4-β-D-mannan with 1,6-α-D-galactose substituents) and konjac gum (1,4-β-D-glucomannan). These gums are remarkable for the synergistic effect they have on other polysaccharides and proteins and their tendency to self-associate. Locust bean gum is also called carob gum.

Guar gum (Hui and Neukom, 1964; Sprenger, 1990) is a shear-stable and hence cold-water-dispersible 1,4-β-D-mannose in linear configuration, with 1,6-linked α-D-galactosyl substituents uniformly spaced along the mannan primary chain. The mannose:galactose ratio has been reported at 1.55 (Morris, 1990) to 1.7 (Grasdalen and Painter, 1980). Locust bean gum (Hui and Neukom, 1964; Sprenger, 1990) differs chemically from guar gum in the higher ratio of mannose to galactose, reported as 3.5 (fewer galactose substituents) (Morris, 1990), and in having a higher DP. Galactose is distributed in copolymer blocks of substituted and unsubstituted mannans. Many more unsubstituted regions exist in the primary structure of locust bean gum than in guar gum, and as a result of the higher mannose content and fewer branch points, this biopolymer is less hydratable than guar gum. However, when heated to 80°C, locust bean gum suspensions become transformed to highly viscous dispersions. α-Galactosidase removes a higher percentage of galactose from locust bean gum than from guar gum. Only about 20% of guar gum can be hydrolyzed (Hui and Neukom, 1964).

Over a solute concentration of 1%, galactomannan pastes have the appearance of a gel. The dispersed solute is mildly pH sensitive between pH 4 and 8; above pH 8, there is a precipitous decline in viscosity. Aqueous dispersions tolerate 10% ethanol. Locust bean and guar gums are good stabilizers of oil-in-water emulsions (Reichman and Garti, 1991), but excessively low concentrations of either may actually destabilize them (Dickinson and Galazka, 1991). A popular use of guar gum is in ice cream manufacture, where allegedly it enhances the smooth texture and slow meltdown property.

Konjac gum contains approximately 85% glucomannan in 1,4-β-D linkage in a glucose:mannose ratio of 1:1.6, and 3–4% acetyl ester groups: there is evidence of branching (Kato and Matsuda, 1973). At 25°C, viscosity is about twice the value for guar and locust bean gums at the same concentration (FMC, 1989). The crystallization tendency is encumbered by the acetyl groups: deacetylation (by mild alkali) results in ordering (Tye, 1991) and consequently in a strong, elastic, thermostable gel at concentrations in excess of 1%; higher concentrations yield an irreversible, rubbery gel. Konjac dispersions increase in gel strength after freezing and thawing (Nakao *et al.*, 1991).

Konjac flour gum is reported to be pH- and cation- (sodium, potassium, and calcium ions) insensitive; this is consistent with its nonionic character (FMC, 1989), but an isolate did not show a characteristic linear viscosity–concentration profile in water; it did show linearity in the presence of electrolytes (Jacon *et al.*, 1993). The apparent partial specific volume of a 0.2–0.4% dispersion was constant over a wide pH range and increased with increasing temperature from 5 to 50°C; it then remained constant (Kohyama and Nishinari, 1993).

C. Fructans

Fructose is the repeating monomer in this class of biopolymers, represented by inulin found significantly in dahlia tubers, chicory root, onions, garlic, and Jerusalem artichoke in which the dry matter approximates 60%. All indications are that inulin is a nonreducing, nongelling, hygroscopic fructofuranose containing 1,2-β- and 2,6-β-D-glycosidic bonds.

In water, inulin undergoes reversion from a more soluble to a less soluble form, in the manner of retrograding starch (Whistler and Smart, 1953): it is slightly soluble in organic solvents. Fructans are easily hydrolyzed by acid. Levan is the branched isomer of inulin.

D. Glycuronans

The chemistry of these polysaccharides is dominated by partial esterification of the total number of carboxyl groups. The sequence of uronic acids in the primary structure is occasionally interrupted by rhamnose, and there is usually a trace of acetyl and phenolic substituents. Models developed from experimental data considered neutral side chains with DP = 2–10 (De Vries *et al.*, 1982). These 1,4-α-linked linear uronans are susceptible to alkali—more so if the C-6 hydroxyl is esterified, but exceptionally acid-stable when this site is unsubstituted. Dispersion stability is less at higher DP. Uronans are endowed with strong dye-fixing and mineral-sequestering properties because of their charged surface.

1. The Pectic Substances

Pectin (unrelated to amylopectin) is the collective name of the galacturonans that are capable of gelling with water, sugar, acid, and/or calcium. High-methoxyl (HM) pectin has DE > 40–50% and low-methoxyl (LM) pectin has DE < 50–40%. HM pectin is the industrial precursor of LM pectin. Demethylation is effected by chemicals or enzymes. Ultrasonication improves the yield of deesterified pectin (Panchev *et al.*, 1994). Completely demethylated pectin is pectic acid. In the wholly protonated form (strongly acidic media), pectic acid quickly precipitates from solution.

Aqueous dispersions of HM and LM pectin have very low viscosity (Walter *et al.*, 1985) and tolerate moderate amounts of ethanol (Walter and Sherman, 1983). HM pectin is unique among hydrocolloids in its ability to gel in acidic media (pH < 3) amid a high concentration of sugar (65%). The gelation mechanism is uncertain, but is known to involve stiffening of the primary chains through protonation, water inactivation by high soluble–solids content, and network coupling. LM pectin requires less sugar and gels by cooperative association through calcium bridges. In juice–milk beverages, HM pectin is preferable to LM pectin, because of its insensitivity to Ca^{2+}. HM pectin jelly is not ordinarily heat-reversible, whereas LM pectin jelly is. HM pectin gels prepared at no higher than 50°C are metastable (Walter and Sherman, 1986).

Gelation of sugar beet pectin is a coupling reaction of feruloyl groups with some oxidants in a way that releases free radicals (Thibault *et al.*, 1991).

2. Alginate

Alginate (Cottrell and Kovacs, 1980; Sime, 1990) is a linear, heteropolysaccharide consisting of 1,4-β-D-mannuronans and 1,4-α-L-guluronans (alginic acid) in variable amounts, depending on the source. Three blocks of copolymers have been identified—a mannuronan sequence, a guluronan sequence, and an alternating mannuronan–guluronan sequence. None of naturally occurring algin, alginic acid, or alginate of di- and multivalent cations is dispersible in water, but each swells to a pastelike consistency. Sodium alginate is hot- and cold-water-dispersible, nongelling, and capable of tolerating 30–40% ethanol with proportionately increasing viscosity. Sodium alginate is remarkably freeze–thaw stable, which ensures its application in refrigerated foods. A wide spectrum of properties is possible by manipulating the ratio of Na^{+} to Ca^{2+} in mixed salts.

The gelling mechanism of alginate is a cooperative association of double helices of guluronate moieties with Ca^{2+} (Rees, 1969). Concentration and the manner of addition of Ca^{2+} have an influence on gel texture. Alginates with a high guluronan content make strong, brittle, heat-stable gels: mannuronan gels are weak, elastic, and less heat-stable (Kelco, 1986). Calcium alginate gels are thermally irreversible and dimensionally stable in the manner of a covalent network (Cottrell and Kovacs, 1980).

Propyleneglycol alginates do not precipitate from acidic media and their acid stability increases in proportion to the propyleneglycol content. Additionally, those esters with DS > 60% are not precipitated by Ca^{2+}, because the bulky propyleneglycol substituent hinders the prerequisite alignment for gelation and precipitation. These esters are more tolerant of alcohol than the unesterified alginate.

3. Gum Arabic

Gum arabic (gum acacia) (Whistler, 1993) is a slightly acidic, highly branched, complex glucuronan containing a 1,3-β-D-galactan main chain and side chains of 1,6-galactopyranose that are themselves substituted with rhamnose, arabinose, and glucuronic acid. This gum is ordinarily water-soluble to approximately 50% concentration, tolerant of ethanol to approximately 60%, incompatible with most other organic solvents, and nongelling. The solution viscosity is very low, even at 40% solute concentration; for this reason, gum arabic dispersions exhibit mostly Newtonian rheology. The dispersions are strongly acid-sensitive over a wide pH range. The dry gum, heated to 170°C and immersed in water, swells without dissolution to a nonsticky gel (Meer, 1980b). The excellent film-forming characteristics of gum arabic make it a favorite encapsulating polysaccharide.

Gum arabic is covalently bonded to approximately 2% protein that is believed to confer exceptional emulsifying properties (Dickinson and Euston, 1991a; Randall *et al.*, 1988). Anderson (1988) located varying proportions of amino acids at the periphery and in the interior of the complex.

4. Gum Karaya

This gum (Meer, 1980c; Kubal and Gralén, 1948) is a processed, partially acetylated galacturonan exudate characterized by strong cohesiveness and adhesiveness, high acid stability, and the usual polyanion sensitivity. Galactose, glucuronic acid, and rhamnose are minor constituents (Dziezak, 1991).

5. Gum Tragacanth

This uronan (Stauffer, 1980; Whistler, 1993) contains a major neutral, insoluble albeit swellable component called bassorin and a minor water-dispersible, acidic component called tragacanthin. It is one of the more acid- and heat-stable, surfactant polysaccharides, and it has wide application in a miscellany of industries. The primary structures appear to be arabinogalactans and arabinogalacturonan methyl esters incorporating lesser quantities of xylose, rhamnose, and fucose.

Fractionation of the major and minor components of gum tragacanth is achieved by a variety of techniques, e.g., filtration, centrifugation, fractional precipitation, etc. In this last technique, tragacanthin disperses in 3:1 ethanol:water, while bassorin simultaneously precipitates as a gel. Acid hydrolysis of tragacanthin yields tragacanthic acid (Belitz and Grosch, 1987).

6. *Xanthan*

Xanthan (Kelco, 1976; Sandford, 1979) is constructed with a repeating five-member unit of three monosaccharides, viz., β-D-glucose, β-D-mannose, and glucuronic acid in a 2:2:1 ratio. The main chain is a 1,4-β-D-glucan, and the side-chains are trisaccharides of α- and β-D-mannose and glucuronic acid. Acetyl and pyruvate groups are present, but appear to have no role in functionality (Callet *et al.*, 1988). Different concentrations of pyruvic acid in the molecules contribute to different rheologies. Xanthan tolerates 50–60% ethanol, once the ethanol is added after the gum has been dispersed in water; it is hot- and cold-water-dispersible, and disperses directly in glycerol at elevated temperatures (> 65°C). NaCl has little or no effect on aqueous dispersions, except below approximately 0.01% when electroviscosity begins to show. This polysaccharide is exceptionally heat-, acid-, alkali-, enzyme-, shear-, and electrolyte-stable: these stabilities are rooted in the interaction between the main and side chains. Importantly, xanthan develops a highly viscous paste that simulates a weak elastic gel as the dispersed solute gradually loses mobility (cooling) without initiating a phase change. Xanthan dispersions have an identifiable τ_0 that is of paramount importance in suspension stabilization. Its exceptional acid stability makes it the stabilizer of choice (over most other glycuronans) for inclusion in fruit juice drinks and vinegar-based dressings.

7. *Gellan*

Gellan gum (Sanderson, 1990) is a partly acetylated linear heteropolysaccharide whose repeating unit is a tetramer of one 1,3-β-D-glucose, one 1,4-β-D-glucose, one 1,3-β-D-glucuronic acid, and one 1,4-α-L-rhamnose. Gellan structure and conformation are conducive to crystallite formation (Chandrasekaran *et al.*, 1988b). Functional properties of commercial gellan gum depend on the acetyl content: partial deacetylation is necessary for water dispersibility. This property is augmented both by a low ionic environment and a total conversion of the mixed-cation extract to the monovalent salt. The partially deacylated fibrils behave somewhat as starch granules: they do not disperse in water at low temperatures (below approximately 70°C), but do at high temperatures. This polysaccharide is a starch mimetic, given its partial cold-water insolubility and transformation of a suspension to a highly viscous dispersion upon heating (Sanderson, 1990). Viscosity is increased significantly by electrolytes to a state of gelation at room temperature at as low a gellan concentration as hundreths of a percent: a concentration of 0.5% yields the firmest gel. Gellan gels are soft and elastic in low ionic-strength media and firm and brittle in high ionic strength media. The sols show a temperature hysteresis, which develops upon cooling from above 70°C to 20–50°C, but no melting until 65–125°C. Thermal reversibility and irreversibility are inducible by controlling the cation concentration.

Robinson *et al.* (1991) schematically represented a weak gellan gel held together by helical associations in the absence of cations and a strong gel assembled by cation-mediated aggregates of helices.

8. Hyaluronic Acid

Although not currently a recognized food polysaccharide, hyaluronic acid (Whistler and Smart, 1953) is of interest, inasmuch as it is the strongest of the linearly configured uronans: it is composed of equimolar quantities of *N*-acetylglucosamine and glucuronic acid. Its importance in living animal tissues is in its lubricity, bonding, and transport properties. Hyaluronic acid complexes with protein and precipitates under acidic conditions.

E. Sulfated Glycans

Sulfate in this class confers much greater acidity and dispersion stability than does the carboxyl group in glycuronans.

1. Agar

Agar presents an interesting example of a mixed polysaccharide stabilizer containing variable quantities of a nongelling uronan and a gelling glucan (agarose; also called agaran). This biopolymer is a linear heterogalactan with 1,3-, 1,4-, and 3,6-α- and β-anhydroglycoside linkages. Agaran is neutral and helical with a polar interior (Morris and Norton, 1983). Reportedly it undergoes retrogradation similarly to amylose when isolated (Hayashi and Kanzaki, 1987). Agaropectin is the other major component; it has a low degree of sulfation and small quantities of pyruvic and uronic acids.

Agar is cold-water-insoluble, but hot-water-soluble. Its sols are heat-stable with essentially constant viscosity in the neutral pH range, and it is self-gelling at as low as hundreths of a percent; higher sulfation increases the critical gelling concentration (Rees, 1972c). Agar dispersions exhibit temperature hysteresis, whereby a 1.5% heated aqueous sol gels at 32–39°C upon cooling and does not solate until heated again to 60–90°C. T_m is raised by increasing the concentration; salt hastens the onset of gelation; sugar increases gel firmness and enhances gel transparency (Matsuhashi, 1990);

when cooled to or below 0°C, gel properties are lost and the solute precipitates.

Agar gels are distinguished from most other hydrocolloids by the sharp, rigid boundary they share with water at an interface, upon cooling a hot sol (Fig. 1 in Chapter 2). Consequently, this biopolymer has been applied on occasion to antisloughing and to sealing cut fruit and vegetable surfaces. Agar precipitates tannins from wine, and its superb stability to enzymes accounts for its universal use in solidifying microbiological culture media.

2. *Carrageenans*

Carrageenans (Moirano, 1977; Guiseley *et al.*, 1980; Yalpani, 1988; Therkelsen, 1993) are linear heteropolysaccharides structurally related to agar, but have a higher sulfate content. The main chain consists of alternating copolymers of 1,4-α- and 1,3-β-D-galactopyranose and 3,6-anhydro-D-galactopyranose: pyruvate and methoxyl groups are minor substituents. The natural extract is a mixed salt of calcium, sodium, potassium, and magnesium. Partial desulfation with alkali simultaneously replaces calcium and magnesium with sodium and potassium, and increases hydratability and the gelation tendency.

For convenience, the numerous possibilities for sulfation and desulfation within the family of carrageenans are identified by Greek prefixes. The cationic nature and content, the preponderance of glycoside bonding, and the sulfate distributions determine whether the isomer is κ-, λ-, or ι-carrageenan; each form has different sol–gel characteristics. In sols, the ordinarily gelling κ- and ι-carrageenan aggregate through cation mediation (Rees, 1969; Dalgleish and Morris, 1988). κ-Carrageenan gels are firm, brittle, and given to syneresis, but ι-carrageenan gels are soft, elastic, and less prone to this defect (Roesen, 1992). Thermoplasticity is based on the reversibility of the double helices. λ-Carrageenan is cold-water-dispersible and nongelling in any cationic form. In this structure, the C-2 sulfate performs as a wedge to prevent development of a double helix (Moirano, 1977). Sulfate in κ- and ι-carrageenan does not interfere with the orientation, and the double helix is therefore not forbidden; so, on cooling, double helices develop into the building blocks of a three-dimensional network (Rees, 1972a, b, c). Of these three gums, κ-carrageenan is most sensitive to electrolytes; all tolerate large quantities of sugar in the order κ-, λ-, and ι-carrageenan; acid stability is $\iota > \lambda > \kappa$.

The κ- and ι-carrageenan gels show a temperature hysteresis between gelation and melting. The sol–gel transition temperature depends more on the specific ion and its concentration than on the carrageenan. Gels may be formed from carrageenan sols without heat (cold process) by permitting Ca^{2+} to diffuse into the sol. The higher T_{gel} of ι-carrageenan has favored its use in the manufacture of food gels that need not be refrigerated.

Carrageenans complex with protein under conditions where carboxylated polysaccharides do not (Tolstoguzov, 1986). One of their major applications is to stabilize casein in evaporated milk and dairy foods.

3. Furcellaran

Furcellaran is a polysaccharide related to κ-carrageenan; it differs mainly in origin and a smaller quantity and narrower distribution of sulfate.

III. Summary

With few exceptions, all classes of polysaccharides have common heterogeneities that exist across chemical and functional boundaries. In many instances these heterogeneities preclude definitive structure–function correlations. This notwithstanding, origin, fine structure, etc., can sometimes confer subtle distinctions, so that one polysaccharide may be preferable to others for a specific purpose. Dispersions of wheat starch have lower viscosity than dispersions of corn starch at identical concentrations, whereas waxy maize starch is the bodying agent of choice over corn starch, when concern for clarity is secondary to dispersion stability.

Polysaccharides of different origins can have similar structures, and polysaccharides with a common origin can have different structures. Classifying them chemically is the least ambiguous system of characterization, but information gleaned therefrom does not inform that guar gum, for example, is cold-water-soluble, but locust bean gum is not, that agar and gellan show similar temperature hysteresis at different concentrations of solute and electrolytes, that low-acyl gellan gum is texturally similar to agar and κ-carrageenan and can replace them (Sanderson *et al.*, 1988), and that agar gels behave similarly to starch gels (Oates *et al.*, 1993). Neither does chemical classification suggest a relatively hydrophobic interface between water and ionic agar, alginate and acacia gums, and nonionic konjac. There are hardly recognizable differences between ionic and nonionic polysaccharides where electrolyte sensitivity is not a factor. Moreover, different combinations of stimuli can elicit the same response in different dispersions and, conversely, the same stimulus can elicit a different response in the same class.

Regardless of classification, polysaccharides express varying degrees of ethanol tolerance: neutral species are mostly unresponsive to electrolytes, including weak acid, and protonated species precipitate with low solvent retention from strong acid and Ca^{2+} solutions. Exceptions to many general rules abound; for example, polyacids are generally the most stable species, but the C-6 methyl ester uronan—pectin—is unusually more labile than the C-6 alginate ester uronan—propyleneglycol alginate. Xanthan expands but CMC contracts in dilute electrolyte solutions, after viscoelasticity. Given all the commonalities and criticalities, dispersed polysaccharides can logically be assumed to be functionally interchangeable, irrespective of any systematic grouping.

CHAPTER **9**

Saccharides in Fat Replacement

I. Introduction

Supplementing cellulose, starch, pectin, and the exudate gums, many intermediate DP (100–200) saccharides comprise a sizable fraction of the cell-wall matrix of a terrestrial plant. The concentrations are highest in the seed coats of grain, parenchymatous stems, and the pericarp of fruits and vegetables. These saccharides also display hydrocolloidal activity, are nongelling, and are immune to the action of human digestive enzymes. They are to cereal grain what cellulose is to vegetative tissue. Between the extremes of a simple sugar and a polysaccharide, they span a wide range of $\overline{M}$ and a correspondingly wide distribution of properties: for example, concurrently with their ability to scatter light, a characteristic of polymers, they exhibit strong reducing power. Some properties, e.g., dispersion, transparency, and stability, are superior to those of polymers.

A. Hemicellulose

The hemicelluloses are polymolecular, random-coil, often branched heteroglycans interspersed with cellulose microfibrils and starch granules *in vivo*. Glucose, xylose, and arabinose are the prominent sugars, although lesser quantities of galactan, mannan, etc., and minor concentrations of acetic, ferulic and uronic acids, etc., are in evidence. Heterogeneity extends to dimensions and conformation.

The total hemicellulose content of plant materials is 15–30% (Norman, 1943), depending on species, cultural practices, and maturation. Collectively, they are second only to celluose in natural abundance.

The noncellulose β-D-glucans in hemicellulose are generally 1,3- and 1,4-bonded arabinans, xylans, arabinoxylans, arabinogalactans, etc. Their aqueous dispersions are quite viscous to the detriment of filters used to clarify cereal-based fermented beverages. Oat β-glucans are rheologically similar to guar gum (Autio, 1996). The pentosans and glucans in hemicellulose, by themselves or through association with residual protein, are intensely surfactant: those in cereal flour containing traces of ferulic acid undergo oxidative gelation (Hoseney, 1986).

B. Oligosaccharides

Oligosaccharides are intermediate DP hydrolysates of polysaccharides or are synthesized by enzymes. They are empirically differentiated from polysaccharides by a chain length of 2–20 monomers (Munk, 1989); Ikan (1991) limits the differentiation to ten monomers. These species are important moieties in polysaccharide structural elucidation by gas chromatography and mass spectroscopy, because their long segments are identical to those of the original primary chain. By manipulating the reaction variables, the DP can be tailored to functional specifications.

Commercial dextrins are specifically the oligomers of starch. White dextrins, so called because of their visual appearance, are produced from a 30–40% suspension under the mildest possible hydrolysis conditions (79–120°C for 3–8 h in 0.2–2% H_2SO_4 or HCl). Yellow dextrins and British gums are the partial hydrolysates at higher time–temperature integrals. Maltodextrins, dextrose equivalent[20] 5–19, derive from controlled enzyme or acid partial hydrolysis of gelatinized corn starch. The 20–24 dextrose equivalent hydrolysates are corn syrups (Appi, 1991).

Amylodextrin is the homogeneous product of prolonged hydrolysis of starch below T_{gz}, terminating in the crystalline equivalent of Avicel and approximating 25 glucose monomers. The hydrolysis is normally at room temperature over intervals of months, wherein the amorphous regions are degraded and the starch crystallites are left intact. Unlike Avicel crystallites, starch crystallites may be disrupted by stress (Kerr, 1950).

Cyclodextrins (Parrish, 1987; Szejtli, 1981, 1991) are water-soluble, cyclic oligosaccharides (cycloamyloses) formed by the action of glycosyl transferase (from *Bacillus macerans*) on amylose. The designation, α, β, and γ (Schardinger compounds) indicates ring formation with six, seven, and eight glucopyranose sections, respectively, in ascending order of cavity size. β-Cyclodextrin is currently the industrially important oligomer of the three. The existence of "tight helical regions" in amylose has not gone undisputed (Banks and Greenwood, 1975), but the ease of the biocatalytic cyclization

20. Dextrose equivalent equals 100 times the weight of reducing sugar as dextrose per unit weight of dry substance.

reaction gives credence to the concept of the linear segments of starch in helical conformation (Szejtli, 1991). As the product of specific enzyme action, the cyclodextrins are recovered from water in 99% purity.

Polydextrose (Appi, 1991) is a synthetic, randomly bonded, amorphous, condensation heterooligomer of glucose, sorbitol, and citric acid, with $\overline{M}$ not exceeding 5000 Da and a solution pH of 2.5–3.5. Its solution viscosity is slightly higher than that of sucrose at similar concentrations, and it forms a clear melt above 130°C.

The partial hydrolysis of fructans yields fructooligosaccharides. *Aspergillus niger*, a fungus, produces fructooligosaccharides in commercial quantities from sucrose. These oligomers, claimed to be 0.4–0.6 times as sweet as sucrose, are indigestible by humans (Spiegel *et al.*, 1994).

II. Isolation

Hemicellulose and oligosaccharides are characterized according to their water and alkali solubility: branching and density are the major distinguishing features. Where the lignin content is low, e.g., in the endosperm of cereals, hot water extracts a considerable percentage.

In a typical hemicellulose isolation process, finely ground plant material (200 g) is treated with 5–10% sodium hydroxide in oxygen-free water at 25°C for 18–24 h (Whistler and Feather, 1965). The water-insoluble fraction, containing two subfractions differentiated by density, is extractable with higher rather than lower NaOH concentrations: progressively higher concentrations of alkali extract progressively more of the dense saccharides.

Considering the heterogeneities of the intermediate DP saccharides (composition, bonding, conformation, DP, etc.), there is no single criterion of purity, but to the extent that extracts can be considered pure, the highest grades are obtained from holocellulose (i.e., cellulose previously treated to remove lipids, starch, pectin, water-soluble carbohydrates, and lignin). Acidification of the alkali supernatant liquid results in a precipitate, hemicellulose A, that is rich in the higher DP water-insoluble xylans and poor in arabinans and uronans. The acidic supernatant liquid retains hemicellulose B, that in contrast to hemicellulose A, has lower DP, higher concentrations of arabinan and glucuronans, and branching (Dekker, 1979; Hoseney, 1986). Hemicellulose B is recovered from the acidic supernatant liquid by precipitation with ethanol. In another fractionation process with water–sodium carbonate–butanol, hemicellulose was separated in the aqueous phase and lignin was separated in butanol, while cellulose remained inert (Magee and Kosaric, 1985).

III. Reactivity

Extraction, purification, and bleaching (with hydrogen peroxide) leave hemicellulose fibers softened, swollen, and with greatly enhanced amorphism and adhesiveness, but with water retentivity (Gould *et al.*, 1990) inferior to polysaccharides, although still quite high. The stability of hemicellulose dispersions is maximum at neutral pH. Dried fibers are difficult to redisperse.

The isolated, intermediate DP saccharides are more kinetically mobile than the polysaccharides, because of their smaller size, and they are more chemically reactive, because of the lower ratio of monomer to reducing end groups. Mild acid hydrolysis splits simple sugars from the pentosan primary structure that, on boiling in concentrated HCl, dismutes to 2-furaldehyde (furfural) in three stages—a primary random hydrolysis, followed by monomerization, and finally monomer decomposition. Knirel *et al.* (1989) used hydrogen fluoride to circumvent certain problems inherent in the dismutation; they claimed that high and changing specificities were a function of reaction temperature (below freezing).

The reaction of hemicellulose to heated, concentrated (12%) HCl is a test of the former's presence. Detection is by the violet or red color developed by one or more of the reaction products (chiefly furfural) with orcinol or phloroglucinol. The coloration is either blue, greenish, or nonexistent with $KI \cdot I_2$ (Whistler and Smart, 1953). The quantity of furfural produced relates directly to the quantity of pentosan present (Smith and Montgomery, 1959). Under the same reaction conditions, hexoses produce hydroxymethylfurfural. Furfural is steam-distillable; hydroxymethylfurfural is not.

Hemicellulose immersed in alkali saponifes, enolizes, deesterifies, depolymerizes by β-elimination, and oxidizes in the presence of air.

IV. Uses

The food, feed, and paper industries use hemicellulase to convert hemicellulose to useful products (Wong and Saddler, 1993). At the concentrations found in vegetable waste matter, hemicellulose has been viewed intermittently as a potential enzyme substrate and as feedstock for commercial ethanol and furfural production.

Pure cellulose makes inferior paper, because cellulose fibers suffer appreciable physical degradation during the pulp beating operation. A certain amount of hemicellulose, especially the uronic acid subfraction, improves quality (Whistler and Smart, 1953).

The reactivity of hemicellulose and oligosaccharides is an asset in food fabrication. As intentional food additives, these intermediate DP saccharides perform numerous functions, e.g., inhibition of sucrose crystallization, surfactancy, clarification, sweetening, binding, antistaling, anticaking, antibrowning (antioxidation), humidification, and the transport of flavor and aroma. They make thin, opalescent dispersions resistant to settling, because the population of crystallites and the DP are low. The water retentivity of β-glucans imparts high viscosity and visco-elasticity to their dispersions, thereby enabling them to retain large volumes of gas in doughs and batters—a property that persists throughout baking (Jeltema *et al.*, 1983). The dense, insoluble subfractions swell or can be sheared into similarly viscous and viscoelastic fluids.

Hemicellulose has a greater impact on food texture than do polysaccharides (Kim and Kim, 1988), because it is this group of natural compounds that differentiates crispiness in vegetables (Klockeman *et al.*, 1991). The many physical changes it undergoes with time directly affect the texture of cooked plant tissues whose quality attributes remain resident more in the low DP than in the high DP component (Kim and Kim, 1988).

One of the arguments for the use of β-cyclodextrin in food is that it can complex deleterious hydrophobic compounds like cholesterol and caffeine whose concentrations may then be lowered or eliminated (Oakenfull and Sidhu, 1992; Hedges *et al.*, 1993). By the same mechanism, β-cyclodextrin can emulsify water–triglyceride mixtures (Shimada *et al.*, 1992), protect light-, heat-, and oxygen-sensitive compounds, eliminate undesirable flavors and hazes, and stabilize desirable ones (Miyawaki and Konno, 1982).

Hemicellulose and oligosaccharides are now staple items in fat replacement systems.[21] They are claimed to have indirect beneficial roles in human health, because they ferment in the intestines (Tomomatsu, 1994) and large bowel where they produce short-chain fatty acids that seem to enhance electrolyte absorption and stimulate colonic muscular activity (Topping, 1994). Polydextrose is used as a low-calorie bulking agent.

V. Fat and Fat Replacement

A diminution of the quantity of fat in foods, without compensating additions, automatically leads to increased concentrations of the other components: for example, a formula consisting of 80% water, 10% protein, 5% carbohydrate, and 5% fat would automatically have a carbohydrate increase from 5 to 5.3% of the mixture (from 25 to 33.3% of the solid phase), if the fat were

21. Fat substitute, mimetic, and replacer are herein used interchangeably.

completely removed. In this sense, water, carbohydrates, and proteins are fat replacers.

A low-fat or fat-free formula cannot have the same gustatory, tactile, textural, and rheological characteristics as a fat-laden formula. Why? Fat contributes uniquely to food through its nonpolar chemical composition, its physical response to low heat (softening), and its interaction with other ingredients. Nevertheless, to fabricate low-fat or fat-free products without appreciable loss of quality, ingredient manufacturers and processors have developed facsimiles of fat, using poly- and oligosaccharides, proteins, and blends. The observations of Campbell (1989) on unsaturated monoglycerides versus saturated glycerides in oil–water stabilization suggest that the less amphiphilic (more ethanol-tolerant) polysaccharides should make better stabilizers and fat replacers.

Jones (1996) recommended a "holistic approach" to fat replacement, whereby the spherical shape of a fat droplet as well as the properties of a fat are mimicked by using hydrophobic compounds. Accordingly, the least amphiphilic saccharides with the shortest persistence lengths, e.g., pectin and the cellulose derivatives (Lapasin and Pricl, 1995), are suitable models. By their nature, the intermediate DP saccharides fulfil the requirement.

A. Essential Roles in Food

A constitutively hydrophobic food fat is called upon to perform in a hydrophilic environment, suggesting the most important physical property to be its emulsifying capacity. Simultaneously, the fat is ingredient-compatible, heat-stable, and confers a smooth oral sensation. The basic assignment of a fat replacer is to mimic these properties through substitution for the fat's viscosity, texture, and the slippery, creamy, lubricious mouthfeel (Glicksman, 1991). One of the earliest fat substitutes to perform thus was Simplesse (Roller and Jones, 1996), a protein perceived to be of a creamy texture due to inherent 0.1–3.0-μm-diameter microparticles (Thayer, 1992).

In dough and batter, fat is thought to waterproof starch particles, thereby preventing hardening of the gluten–starch suspension (Platt and Fleming, 1923). This tenderization process is called shortening and the lipid material (usually a special mixture of glycerides) is the shortener. A characteristic of the shortener is that the phase transition from solid to liquid is completed over a narrow plastic range. The brittle texture of cakes and pastries depends on retention or restoration of a certain percentage of fat crystals after baking and cooling. By supplying the stabilizing crystallites and emulsifying a large volume of air and water in dough and batter (creaming), shortenings ensure that the product has a moist yet friable structure.

Glicksman (1991) organized fat substitutes into 10 general categories, viz., synthetic fat substitutes, synthetic emulsifiers, hydrocolloids, starch derivatives, hemicelluloses, β-glucans, soluble bulking agents, microparticulates, composite materials, and functional blends. Edible chemical compounds consisting of carbon backbones with carboxyl and methylcarboxyl groups as ester and acyl substituents have been disclosed as fat mimetics—some with special advantages (Klemann and Finley, 1989). Saccharides are identifiable with this composition are not ordinarily metabolized by humans, and hence have no calorie content. Gums, maltodextrin, corn syrup, etc., impart "body, cling and firmness" (Yackel and Cox, 1992) in fat replacement.

A fat contains 9.54 kcal g^{-1}, a simple sugar contains approximately one-half that amount, and a polysaccharide is not metabolized; therefore, the calorie contribution of an intermediate DP saccharide fat substitute is proportional to the degree of saccharification.

B. Carbohydrate Fat Mimetics

Carbohydrate fat mimetics (Thayer, 1992) should be thermoreversible gels (e.g., starch) and microreticulated fibers (e.g., cellulose). As gels, they should provide plasticizing moisture, and as fibers, they should create structural rigidity. Concentrations higher than normal use levels are necessary for them to mimic the fatty sensation. They may be dry-blended with the rest of the formula, or suspended in water and sheared separately.

A patented product of oat fiber, called Oatrim, containing up to 10% β-glucans, is considered to have fat mimetic as well as cholesterol-lowering capabilities (Roller and Jones, 1996). Pectin (Mangat, 1992), starch, alginate (Kerner and Ward, 1992), maltodextrin (Inglett and Grisamore, 1991), dried fruits (Silge, 1992), cereal, vegetable fibers (Glicksman, 1991), and enzyme-hydrolyzed rice (Sander, 1992) all have been reported to be fat replacers. One of the suggested advantages of hydrolyzed rice solids over most other carbohydrates is that their cold-water gel does indeed resemble a fat. The protein adjunct in rice and other mimetics of natural origin augments the fatty sensation. Other reported fat substitutes are xanthan contributing appropriate viscosity (Duxbury, 1993), high-temperature curdlan gels reputed to have a high fat-adsorbing capacity (Harada *et al.*, 1993), and polydextrose (LaBell, 1991) functioning also as a sugar substitute. Methylcellulose, hydroxypropylcellulose (Dziezak, 1991; Henderson, 1989), and microcrystalline cellulose (Penichter and McGinley, 1991), while imitating fat in stabilizing creams and toppings, are at the same time barriers to lipid migration in fried foods, while promoting adequate moisture retention in the product.

A single polysaccharide can rarely optimize all the requisites of a fat, so the trend has been toward combinations, e.g., konjac–carrageenan blends in lean meat and sausages (FMC, 1991). Simulation should be more achievable by the use of cosolutes with complementary sol–gel characteristics. The substituted celluloses are hydrated and fully viscous at room temperature, but in the vicinity of 50°C, they phase-separate abruptly. By combining any of the substituted celluloses with a normal gelling polysaccharide, e.g., gellan acting like starch, upon heating, the sol–gel transition of the former would be complemented by the latter's gel–sol transition, which would result in deposition of crystallike, stabilizing substituted cellulose particles embedded in a continuous gellan phase. Protein–polysaccharide (i.e., gelatin–galactomannan) antagonism has created similar fat mimetics, consisting of gelatin particles embedded in a continuous galactomannan phase (Muyldermans, 1993), possessing plastic textures different from those of either antagonist. A guar gum–microcrystalline cellulose blend is a superior fat replacer (FMC, 1993).

Micronization is an optional process that helps to reproduce the fatty sensation afforded by polysaccharides. The 0.2-nm particles in Avicel (FMC, 1993) and the 1–5-nm particles in the rice mimetic (Pszczola, 1991) simulate lipid emulsion rheology as well as lipid oral sensations. The simulation mechanism implicates a weak gel structure and an expansive surface where a large volume of water is immobilized.

VI. Summary

The diverse properties of saccharides have made them the subjects of research and development in the contemporary quest for natural fat substitutes. It is now apparent that there will be no single fat mimetic for all applications; so synergistic and antagonistic interactions between saccharides and with protein show promise for specific applications. The exceptional performance of intermediate DP saccharides, currently of limited industrial value, can fulfil a critical role as they have in the past in other industries.

Appendices

Appendix 1

Unit of Viscosity

$$V_i = kr^4 t\tau/(\eta_i \cdot l)$$

$$\mathrm{cm}^3 = k(\mathrm{cm}^4\ \mathrm{s\ g\ cm})(\mathrm{cm}^2\ \mathrm{s}^2\ \mathrm{cm}\ \eta_i\ \mathrm{cm}^3)^{-1}$$

$$\eta_i = k(\mathrm{cm}^4\ \mathrm{s\ g\ cm})(\mathrm{cm}^3\ \mathrm{cm}^2\ \mathrm{s}^2\ \mathrm{cm})^{-1}$$

$$\eta_i = \mathrm{kg\ cm}^{-1}\ \mathrm{s}^{-1}$$

$$V_i = \mathrm{cm}^3$$

$$r = \mathrm{cm}$$

$$t = \mathrm{s}$$

$$\tau = \mathrm{dyn\ A}^{-1} = \left[\mathrm{g\ cm}(\mathrm{cm}^2\ \mathrm{s}^2)^{-1}\right]$$

Appendix 2

The Schulz–Blaschke Equation

$$\eta_{\mathrm{sp}}/c = [\eta] + k'''[\eta]\eta_{\mathrm{sp}}$$

$$= [\eta](1 + k'''\eta_{\mathrm{sp}})$$

$$[\eta] = \{\eta_{\mathrm{sp}}/c]/[1 + k'''\eta_{\mathrm{sp}}\}$$

$$= \eta_{\mathrm{sp}}/(1 + k'''\eta_{\mathrm{sp}})c$$

$$[\eta]^{-1} = \{(1 + k'''\eta_{\mathrm{sp}})/\eta_{\mathrm{sp}}\}c$$

Appendix 3

The Maxwell Model

$\epsilon_{tot} = \tau/G$ (Hooke's law)

$(\epsilon_{tot}/t = \tau/\eta)$ (Newton's law)

$\epsilon_{tot} = (\tau/\eta)t$

$\epsilon_{tot} = \tau/G + (\tau/\eta)t$

$\partial\epsilon_{tot}/\partial t = (1/G)(\partial\tau/\partial t) + \tau/\eta$

$(1/G)\,\partial\tau_\lambda/\partial t = (\partial\epsilon_{tot}/\partial t) - (\tau/\eta)$

$\partial\tau/\partial t = G[(\partial\epsilon_{tot}/\partial t) - (\tau/\eta)]$

ϵ = shear
t = time
τ = stress
G = modulus of elasticity
η = viscosity

Appendix 4

Unit of η/G

From Newton's equation,

$$\begin{aligned} \eta &\approx \text{dyn s cm}^{-2} \\ &\approx \text{g cm s}^{-2}\text{ s cm}^{-2} \\ &\approx \text{g cm}^{-1}\text{ s}^{-1}. \end{aligned} \tag{A4.1}$$

From Hooke's equation,

$$\begin{aligned} G &\approx \text{dyn cm}^{-2} \\ &\approx \text{g cm}(\text{cm}^2\text{ s}^2)^{-1} \\ &\approx \text{g cm}^{-1}\text{ s}^{-2}. \end{aligned} \tag{A4.2}$$

By rationalizing A5.1 and A5.2

$$\begin{aligned} \eta/G &\approx \text{g cm}^{-1}\text{ s}^{-1}(\text{g cm}^{-1}\text{ s}^{-2})^{-1} \\ &\approx \text{s}. \end{aligned}$$

Appendix 5

The Voigt–Kelvin Model

$\tau = G\epsilon$ (Hooke's equation)

$\epsilon = \tau/G$

$\tau = \eta\epsilon/t$ (Newton's equation)

$\epsilon = \tau t/\eta$

$\tau = \eta\epsilon/t + G\epsilon$

$\tau - G\epsilon = \eta\epsilon/t$

$\epsilon/t = (\tau - G\epsilon)/\eta$

$\epsilon/t = \tau/\eta - G\epsilon/\eta$

$\partial\epsilon/\partial t = \tau/\eta - G\epsilon/\eta$

$\epsilon = (\tau/\eta - G\epsilon/\eta)t$

ϵ = shear

t = time

τ = stress

G = modulus of elasticity

η = viscosity

Appendix 6

The Mark–Houwink Equation and the Hydrodynamic Volume

$[\eta]_1 M_1 = [\eta]_2 M_2$

$\log[\eta]_1 + \log M_1 = \log[\eta]_2 + \log M_2$

$[\eta] = KM^{\nu}$

$\log[\eta] = \log K + \nu \log M$

Substituting $[\eta]$ from Eq. (A6.4) for $[\eta]_1$ and $[\eta]_2$ in Eq. (A6.2)

$$\log K_1 + \nu_1 \log M_1 + \log M_1 = \log K_2 + \nu_2 \log M_2 + \log M_2$$

$$\log K_1 + (\nu_1 + 1)\log M_1 = (\nu_2 + 1)\log M_2 + \log K_2$$

$$\log M_2 = \{(\nu_1 + 1)\log M_1 + \log K_1 - \log K_2\}/(\nu_2 + 1)$$

$$\log M_2(\nu_2 + 1) = (\nu_1 + 1)\log M_1 + \log(K_1/K_2)$$

$$\begin{aligned} \log M_2 &= \{(\nu_1 + 1)\log M_1 + \log(K_1/K_2)\}/(\nu_2 + 1) \\ &= \{(\nu_1 + 1)\log M_1\}/(\nu_2 + 1) + \log(K_1/K_2)/(\nu_2 + 1) \\ &= \{(\nu_1 + 1)\log M_1\}/(\nu_2 + 1) + (1/(\nu_2 + 1))(\log(K_1/K_2)). \end{aligned}$$

When $\nu_1 = \nu_2$ and $K_1 = K_2$,

$$\log M_2 = \log M_1 .$$

References

Adamson, A. W. (1990). *Physical Chemistry of Surfaces*, 5th ed., Interscience, New York.

Aguilera, J. M. (1992). Generation of engineered structures in gels. In *Physical Chemistry of Foods*, Schwartzberg, H. G., and Hartel, R. W. (Eds.), pp. 387–421, Dekker, New York.

Albersheim, P., Neukom, H., and Deuel, H. (1960). Splitting of pectin chain molecules in neutral solution. *Arch. Biochem. Biophys.* 90:46–51.

Alberty, R. A., and Silbey, R. J. (1992). *Physical Chemistry*. Wiley, New York.

Alfrey, T. (1947). The influence of solvent composition on the specific viscosities of polymer solutions. *J. Colloid Sci.* 2:99–114.

Alfrey, T., Bartovics, A., and Mark, H. (1942). The effect of temperature and solvent type on the intrinsic viscosity of high polymer solutions. *J. Am. Chem. Soc.* 64:1557–1560.

Allcock, H. R., and Lampe, F. W. (1981). *Contemporary Polymer Chemistry*. Prentice-Hall, Englewood Cliffs, NJ.

Allen, J. C., Gardner, R., Wedlock, D. J., and Phillips, G. O. (1982). The use of 2-thiobarbituric acid in the assay of κ-carrageenan in polysaccharide mixtures. *Prog. Food Nutr. Sci.* 6:387–392.

Anderegg, R. J. (1990). Mass spectrometry: an introduction. In *Methods of Biochemical Analysis: Biomedical Applications of Mass Spectrometry*, Suelter, C. H., and Watson, J. T. (Eds.), Vol. 34, pp. 1–89. Interscience, New York.

Anderson, D. M. W. (1988). The structural significance of amino acids in some plant gums. In *Gums and Stabilizers for the Food Industry*, Phillips, G. O., Williams, P. A., and Wedlock, D. J. (Eds.), Vol. 4, p. 31. IRL Press, Oxford.

Anderson, D. M. W., Howlett, J. F., and McNab, C. G. A. (1986). The amino acid composition of the proteinaceous components of konjac mannan, seed endosperm galactomannans and xanthan gum. *Food Hydrocoll.* 1:95–99.

Anderson, N. S., Campbell, J. W., Harding, M. M., Rees, D. A., and Samuel, J. W. B. (1969). X-ray diffraction studies of polysaccharide sulfates: double helix models for κ- and ι-carrageenans. *J. Mol. Biol.* 45:85–89.

Anderson, N. S., Dolan, T. C. S., Penman, A., Rees, D. A., Mueller, G. P., Stancioff, D. J., and Stanley, N. F. (1968). Carrageenans. Part IV. Variations in the structure and gel properties of κ-carrageenan, and the characterisation of sulfate esters by infrared spectroscopy. *J. Chem. Soc. C*, Part 1, 602–606.

Anger, H., and Berth, G. (1985). Gel-permeation chromatography of sunflower pectin. *Carbohydr. Polym.* 5:241–250.

Anonymous (1991). Pectin: nature's own fat replacer. *Prepared Foods* 60(12):57.

AOAC (1980). Starch. In *Official Methods of Analysis*, p. 221. Am. Assoc. Offic. Anal. Chem., Washington, DC.

AOAC (1990). Starch in cereals. In *Official Methods of Analysis*, p. 789. Am. Assoc. Offic. Anal. Chem., Washington, DC.

Appi, R. C. (1991). Confectionery ingredients from starch. *Food Technol.* 45:148–149.

Aqualon Company (1987). Klucel Hydroxypropylcellulose. Physical and Chemical Properties. Report 250-2C Rev. 10-87 15M IP. Aqualon Company, Wilmington, DE.

Aries, R. E., Gutteridge, C. S., Laurie, W. A., Boon, J. J., and Eijkel, G. B. (1988). A pyrolysis-mass spectrometry investigation of pectin methylation. *Anal. Chem.* 60:1498–1502.

Arnold, K. M., Blount, W. W., Lawniczak, J. E., Lowman, D. W., and Edgar, K. J. (1994). Synthesis and properties of cellulose acetoacetates. 208th Annual Meeting, Am. Chem. Soc., Div. Cellulose, Paper and Textiles, August 21–26, Washington, DC.

Arnott, S., Fulmer, A., Scott, W. E., Dea, I. C. M., Moorhouse, R., and Rees, D. A. (1974). The agarose double helix and its function in agarose gel structure. *J. Mol. Biol.* 90:269–284.

Aspinall, G. O., and Cottrell, I. W. (1970). Lemon-peel pectin. II. Isolation of homogeneous pectins and examination of some associated polysaccharides. *Can. J. Chem.* 48:1283–1289.

Atkins, E. D. T., Mackie, W., and Smolko, E. E. (1970). Crystalline structures of alginic acids. *Nature (London)* 225:626–628.

Autio, K. (1996). Functional aspects of cereal cell wall polysaccharides. In *Carbohydrates in Food*, Eliasson, A. (Ed.). Dekker, New York.

Axelos, M. A., Lefebvre, J., and Thibault, J. F. (1987). Conformation of a low methoxyl citrus pectin in aqueous solution. *Food Hydrocoll.* 1:569–570.

Badenhuizen, N. P. (1964). General method for starch isolation. *Methods in Carbohydrate Chemistry*, Vol. IV, pp. 14, 15. Academic Press, New York.

Bahary, W. S. (1973). Recent trends in the determination of molecular weights and branching in elastomers. In *Polymer Molecular Weight Methods*, Gould, R. F. (Ed.), pp. 85–97. Am. Chem. Soc., Washington, DC.

Baianu, I. C. (1992a). Molecular structures and chemical bonding. In *Physical Chemistry of Food Processes*, Baianu, I. C. (Ed.), Vol. 1. Van Nostrand–Reinhold, New York.

Baianu, I. C., Ed. (1992b). *Physical Chemistry of Food Processes*, Vol. 1, p. 94. Van Nostrand–Reinhold, New York.

Baianu, I. C., Pessen, H., and Kumosinski, T. F. (1993). *Physical Chemistry of Food Processes*, Vol. 2. Van Nostrand–Reinhold, New York.

Baines, Z. V., and Morris, E. R. (1988). Effect of polysaccharide thickness on organoleptic attributes. In *Gums and Stabilizers for the Food Industry*, Phillips, G. O., Williams, P. A., and Wedlock, D. J. (Eds.), Vol. 4, pp. 193–201. IRL Press, Oxford.

Baines, Z. V., and Morris, E. R. (1989). Suppression of perceived flavor and taste by food hydrocolloids. In *Food Colloids*, Bee, R. D., Richmond, P., and Mingins, J. (Eds.), pp. 184–192. Royal Chem. Soc., London.

Baird, J. K., and Smith, W. W. (1989a). An analytical procedure for gellan gum in food gels. *Food Hydrocoll.* 3:407–411.

Baird, J. K., and Smith, W. W. (1989b). A simple colorimetric method for specific analysis of food-grade galactomannans. *Food Hydrocoll.* 3:413–416.

Baker, T. J., Schroeder, L. R., and Johnson, D. C. (1978). Dissolution of cellulose in polar aprotic solvents via formation of methylol cellulose. *Carbohydr. Res.* 67:C4–C7.

Ball, D. H., Cimecioglu, A. L., Kaplan, D. L., and Huang, S. H. (1994). Site-selective modification of polysaccharides under homogeneous conditions. New anhydro and cationic polymers derived from amylose and pullulan. 208th Annual Meeting, Am. Chem. Soc., Div. Cellulose, Paper and Textiles, August 21–26, Washington, DC.

Balmaceda, E., Rha, C., and Huang, F. (1973). Rheological properties of hydrocolloids. *J. Food Sci.* 38:1169–1173.

Banks, W., and Greenwood, C. T. (1975). *Starch and Its Components*. Edinburgh Univ. Press, Edinburgh.

Barfod, N. M. (1988). Calcium and water-binding activity during alginate gelation. In *Gums and Stabilizers for the Food Industry*, Phillips, G. O., Williams, P. A., and Wedlock, D. J. (Eds.), Vol. 4, p. 285. IRL Press, Oxford.

Barford, R. A., Magidman, P., Phillips, J. G., and Fishman, M. L. (1986). The estimation of degree of methylation of pectin by pyrolysis–gas chromatography. *Anal. Chem.* 58:2576–2578.

Barker, G. C., and Grimson, M. J. (1991). Computer simulations of the flow of deformable particles. In *Food Polymers, Gels and Colloids*, Dickinson, E. (Ed.), pp. 262–271. Royal Chem. Soc., London.

Barker, S. A. (1963). Application of infrared spectra to cellulose. In *Methods in Carbohydrate Chemistry*, Vol. III, pp. 104–108. Academic Press, New York.

Barker, S. A., Bourne, E. J., and Whiffen, D. H. (1956). Use of infrared analysis in the determination of carbohydrate structure. *Methods of Biochemical Analysis*, Vol. 3, pp. 213–245. Interscience, New York.

Barnes, H. A., Hutton, J. F., and Walters, K. (1989). *An Introduction to Rheology*. Elsevier, Amsterdam, New York.

Barth, H. G., (1986). Characterization of water-soluble polymers using size-exclusion chromatography. In *Water-Soluble Polymers*, Glass, J. E., Jr. (Ed.). Advances in Chemistry, 213, pp. 31–55. Am. Chem. Soc., Washington, DC.

Barth, H. G., and Regnier, F. E. (1981). High-performance gel-permeation chromatography of industrial gums: analysis of pectins and water-soluble cellulosics. Methods in Carbohydrate Chemistry, Vol. IX, pp. 105–114. Academic Press, New York.

Barth, H. G., and Smith, D. A. (1981). High-performance size-exclusion chromatography of guar gum. *J. Chromatogr.* 206:410–415.

Barth, H. G., and Sun, S.-T. (1991). Particle size analysis. *Anal. Chem.* 63:1R–10R.

Battista, O. A., and Smith, P. A. (1962). Microcrystalline cellulose. *Ind. Eng. Chem.* 54:20–29.

Bean, J. E., and Bornman, C. H. (1973). Staining of pectic substances as extracts and in the transfusion tracheids of *Welwitschia*. J. S. *African Bot.* 39:23–33.

Bean, M. M., Yamazaki, W. T., and Donelson, D. H. (1978). Wheat starch gelatinization in sugar solutions. II. Fructose, glucose, and sucrose: cake performance. *Cereal Chem.* 55:945–951.

Belitz, H.-D., and Grosch, W. (1989). *Food Chemistry*, pp. 231, 232, 506. Springer-Verlag, New York.

Bellamy, W. D., and Miller, A. A. (1963). The effect of ionizing radiation on cellulose and its derivatives. *Methods in Carbohydrate Chemistry*, Vol. III, pp. 185–189. Academic Press, New York.

Belton, P. S., Wilson, R. H., and Chenery, D. H. (1986). Interaction of group I cations with iota and kappa carrageenans studied by Fourier transform infrared spectroscopy. *Int. J. Biol. Macromol.* 8:247–251.

BeMiller, J. N. (1965a). Alkaline degradation of starch. In *Starch Chemistry and Technology*, Whistler, R. L., and Paschall, E. F. (Eds.), Vol. I, pp. 521–532. Academic Press, New York.

BeMiller, J. N. (1965b). Organic complexes and coordination compounds of carbohydrates. In *Starch Chemistry and Technology*, Whistler. R. L., and Paschall, E. F. (Eds.), Vol. I, pp. 309–329. Academic Press, New York.

Ben-Naim, A. (1980). *Hydrophobic Interactions*. Plenum, New York.

Bergenstahl, B. (1988). Gums as stabilizers for emulsifier covered emulsion droplets. In *Gums and Stabilizers for the Food Industry*, Phillips, G. O., Williams, P. A., and Wedlock, D. J. (Eds.), Vol. 4, pp. 363–369. IRL Press, Oxford.

Berth, G. (1988). Studies on the heterogeneity of citrus pectin by gel-permeation chromatography on Sepharose 2B/Sepharose 4B. *Carbohydr. Polym.* 8:105–117.

Berth, G. (1992), Methodical aspects of characterization of alginate and pectate by light scattering and viscometry coupled with GPC. *Carbohydr. Polym.* 19:1–9.

Berth, G., Anger, H., Plashchina, I. G., Braudo, E. E., and Tolstoguzov, V. B. (1982). Structural study of the solutions of acidic polysaccharides. II. Study of some thermodynamic properties of the dilute pectin solutions with different degrees of esterification. *Carbohydr. Polym.* 2:1–8.

Berth, G., Dautzenberg, H., Lexow, D., and Rother, G. (1990). The determination of molecular weight distribution of pectins by calibrated GPC. Part I. Calibration by light scattering and membrane osmometry. *Carbohydr. Polym.* 12:39–59.

Berth, G., and Lexow, D. (1991). The determination of molecular-weight distribution of pectins by calibrated GPC. II. The universal calibration. *Carbohydr. Polym.* 15:51–65.

Beyer, G. L. (1959). Determination of particle size and molecular weight. In *Technique of Organic Chemistry*, Weissberger, A. (Ed.), Vol. 1, Part I. Interscience, New York.

Biliaderis, C. G. (1992). Structures and phase transitions of starch in food systems. *Food Technol.* 46:99.

Billmeyer, F. W., Jr. (1984). *Textbook of Polymer Science*, 3rd ed., p. 303. Wiley, New York.

Bio-Rad (1971). *Gel Chromatography*. Bio-Rad Laboratories, Inc., Richmond, CA.

Birdi, K. S. (1993). *Fractals in Chemistry, Geochemistry, and Biophysics*. Plenum, New York.

Bitter, T., and Muir, H. M. (1962). A modified uronic acid carbazole reaction. *Anal. Biochem.* 4:330–334.

Blackwell, J. (1982) The macromolecular organization of cellulose and chitin. In *Cellulose and Other Natural Polymer Systems*, Brown, R. M., Jr. (Ed.), p. 403. Plenum, New York.

Blackwell, J., Kolpak, F. J., and Gardner, K. H. (1977). Structures of native and regenerated celluloses. In *Cellulose Chemistry and Technology*, Arthur, J. C., Jr. (Ed.), ACS Symposium Series 48, pp. 42–55. Am. Chem. Soc., Washington, DC.

Blanshard, J. M. V. (1970). Stabilizers—their structure and properties. *J. Sci. Food Agric.* 21:393–399.

Blanshard, J. M. V., and Mitchell, J. R. (1979). *Polysaccharides in Food*. Butterworth, London.

Bourne, M. C. (1982). *Food Texture and Viscosity*. Academic Press, New York.

Bradley, T. D., and Mitchell, J. R. (1988). The determination of the kinetics of polysaccharide thermal degradation using high temperature viscosity measurements. *Carbohydr. Chem.* 9:257–267.

Braudo, E. E. (1992). Mechanism of galactan gelation. *Food Hydrocoll.* 6:25–43.

Brondsted, H., and Kopecek, J. (1992). pH-sensitive hydrogels. In *Polyelectrolyte Gels*, ACS Symposium Series 480, pp. 287–290. Am. Chem. Soc., Washington, DC.

Brownsey, G. J., Cairns, P., Miles, M. J., and Morris, V. J. (1988). Studies on the mechanism of gelation for xanthan–galactomannan and xanthan–glucomannan mixed gels. In *Gums and Stabilizers for the Food Industry*, Phillips, G. O., Williams, P. A., and Wedlock, D. J. (Eds.), Vol. 4, pp. 157–163. IRL Press, Oxford.

Bryce, D. J., and Greenwood, C. T. (1966). The thermal degradation of starch. Part 6. The pyrolysis of amylomaize starch in the presence of inorganic salts. In *Thermoanalysis of Fibers and Fiber-Forming Polymers*, Applied Polymer Symposium, Vol. 2, pp. 159–173. Interscience, New York.

Burchard, W. (1963). Das viskositatsverhalten von amylose in verschiedenen losungsmitteln. *Makromol. Chem.* 64:110–125.

Burchard, W. (1994). Light scattering techniques. In *Physical Techniques for the Study of Food Biopolymers*, Ross-Murphy, S. B. (Ed.), pp. 151–213. Blackie, Glasgow/London.

Buttkus, H. (1970). Accelerated denaturation of myosin in frozen solution. *J. Food Sci.* 35:558–562.

Byrne, G. A., Gardiner, D., and Holmes, F. H. (1966). The pyrolysis of cellulose and the action of flame-retardants. *J. Appl. Chem.* 16:81–88.

Cabane, B., Wong, K., and Duplessix, R. (1989). *Macromolecules for Control of Distances in Colloidal Dispersions*, ACS Symposium Series, 384, pp. 312–327. Am. Chem. Soc., Washington, DC.

Cairns, P., I'Anson, K. J., and Morris, V. J. (1991). The effect of added sugars on the retrogradation of wheat starch gels by X-ray diffraction. *Food Hydrocoll.* 5:151–153.

Callet, F., Milas, M., and Tinland, B. (1988). Role of xanthan structure on the rheological properties in aqueous solutions. In *Gums and Stabilizers for the Food Industry*, Phillips, G. O., Williams, P. A., and Wedlock, D. J. (Eds.), Vol. 4, pp. 203–210. IRL Press, Oxford.

Campbell, I. J. (1989). The role of fat crystals in emulsion stability. In *Food Colloids*, Bee, R. D., Richmond, P., and Mingins, J. (Eds.), pp. 272–282. Royal Chem. Soc., London.

Cao, Y., Dickinson, E., and Wedlock, D. J. (1990). Creaming and flocculation in emulsions containing polysaccharide. *Food Hydrocoll.* 4:185–195.

Carpenter, D. K., and Westerman, L. (1975). Viscometric methods of studying molecular weight and molecular weight distribution. In *Polymer Molecular Weights*, Slade, P. E., Jr. (Ed.), Vol. 2, p. 388. Dekker, New York.

Carroll, B., and Cheung, H. C. (1964). Determination of amylose. *Methods in Carbohydrate Chemistry*, Vol. IV, pp. 170–174. Academic Press, New York.

Cesāro, A., and Villegas, M. (1996). Proton dissociation of ionic polysaccharides: can the molecular weight be approximated by pH determinations? *Food Hydrocoll.* 10:45–50.

Chandrasekaran, R., Millane, R. P., and Arnott, S. (1988a). Molecular structures of gellan and other industrially important gel forming polysaccharides. In *Gums and Stabilizers for the Food Industry*, Phillips, G. O., Williams, P. A., and Wedlock, D. J. (Eds.), Vol. 4, pp. 183–191. IRL Press, Oxford.

Chandrasekaran, R., Millane, R. P., Arnott, S., and Atkins, E. D. T. (1988b). The crystal structure of gellan gum. *Carbohydr. Res.* 175:1–15.

Chandrasekaran, R., Puigjaner, L. C., Joyce, K. L., and Arnott, S. (1988c). Cation interactions in gellan: an X-ray study of the potassium salt. *Carbohydr. Res.* 181:23–40.

Chester, T. L. (1984). Capillary supercritical-fluid chromatography with flame-ionization detection: reduction of detection artifacts and extension of detectable molecular weight range. *J. Chromatogr.* 299:424–431.

Chester, T. L., and Innis, D. P. (1986). Separation of oligo- and polysaccharides by capillary supercritical fluid chromatography. *J. High Res. Chromatogr. Chromatogr. Commun.* 9: 209–212.

Christianson D. D., Hodge, J. E., Osborne, D., and Detroy, R. W. (1981). Gelatinization of wheat starch as modified by xanthan gum, guar gum, and cellulose gum. *Cereal Chem.* 58:513–517.

Chuma, Y., Uchida, S., and Shemsanga, K. H. H. (1982). Simultaneous measurement of size, surface area, and volume of grains and soybeans. *Trans. Am. Soc. Agric. Eng.* 25:1752–1756.

Clark, A. H. (1992). Gels and gelling. In *Physical Chemistry of Foods*, Schwartzberg, H. G., and Hartel, R. W. (Eds.), p. 268. Dekker, New York.

Clark, A. H., and Ross-Murphy, S. B. (1987). Structural and mechanical properties of biopolymer gels. *Adv. Polym. Sci.* 83:57–192.

Clark, R. (1995). Extensional viscosity in liquid foods. Institute of Food Technology, Annual Meeting, Institute of Food Technology, Chicago, IL. Abstract 37-5.

Clark, R. C. (1988). Viscoelastic response of xanthan gum/guar gum blends. In *Gums and Stabilizers for the Food Industry*, Phillips, G. O., Williams, P. A., and Wedlock, D. J. (Eds.), Vol. 4, pp. 165–172. IRL Press, Oxford.

Clark, R. C. (1992). Extensional viscosity of some food hydrocolloids. In *Gums and Stabilizers for the Food Industry*, Phillips, G. O., Williams, P. A., and Wedlock, D. J. (Eds.), Vol. 6, pp. 73–85. IRL Press, Oxford.

Cleemput, G., Roels, S. P., van Oort, M., Grobet, P. J., and Delcour, J. A. (1993). Heterogeneity in the structure of water-soluble arabinoxylans in European wheat flours of variable bread-making quality. *Cereal Chem.* 70:324–329.

Cleland, R. L., Wang, J. L., and Detweiler, D. M. (1982). Polyelectrolyte properties of sodium hyaluronate. 2. Potentiometric titration of hyaluronic acid. *Macromolecules* 15:386–395.

Coates, M. L., and Wilkins, C. L. (1987). Laser-desorption Fourier transform mass spectra of polysaccharides. *Anal. Chem.* 59:197–200.

Coffey, D. G., Bell, D. A., and Henderson, A. (1995). Cellulose and cellulose derivatives. In *Food Polysaccharides and Their Applications*, Stephen, A. M. (Ed.), pp. 123–153. Marcel Dekker, New York.

Cohen Stuart, M. A., Fleer, G. J., and Bijsterbosch, B. H. (1982). The adsorption of poly(vinyl pyrrolidone) onto silica. I. Adsorbent amount. *J. Colloid Interface Sci.* 90:310–320.

Cooke, D., and Gidley, M. J. (1992). Loss of crystalline and molecular order during starch gelatinization: origin of the enthalpic transition. *Carbohydr. Res.* 227:103–112.

Corbett, W. M. (1963). Purification of cotton cellulose. *Methods in Carbohydrate Chemistry*, Vol. III, pp. 3–9. Academic Press, New York.

Cottrell, I. W., and Kovacs, P. (1980). Alginates. In *Handbook of Water-Soluble Gums and Resins*, Davidson, R. L. (Ed.), pp. 2-1–2-43. McGraw-Hill, New York.

Cottrell, I. W., Kang, K. S., and Kovacs, P. (1980). Xanthan gum. In *Handbook of Water-Soluble Gums and Resins*, Davidson, R. L. (Ed.), pp. 24-7ff. McGraw-Hill, New York.

Cowie, J. M. G. (1973). *Polymers: Chemistry and Physics of Modern Materials*. Intertext Books, London.

Cowie, J. M. G. (1991). *Polymers: Chemistry and Physics of Modern Materials*, 2nd ed. Blackie Glasgow, London.

Crescenzi, V., and Dentini, M. (1988). Solution conformations of the polysaccharide gellan. In *Gums and Stabilizers for the Food Industry*, Phillips, G. O., Williams, P. A., and Wedlock, D. J. (Eds.), Vol. 4, pp. 63–69. IRL Press, Oxford.

Crowe, N. L. (1989). Effect of chlorine compounds and other oxidants on the oxidative gelation and cake baking properties of wheat flour pentosans. *Diss. Abstr. Int. B.* 50:13.

Cui, W., Eskin, N. A. M., and Biliarderis, C. G. (1993). Water-soluble yellow mustard (*Sinapis alba L.*) polysaccharides: partial characterization, molecular size distribution and rheological properties. *Carbohydr. Polym.* 20:215–225.

Dalgleish, D. G., and Morris, E. R. (1988). Interactions between carrageenans and casein micelles: electrophoretic and hydrodynamic properties of the particles. *Food Hydrocoll.* 4:311–320.

Daniels, F., Mathews, J. H., Williams, J. W., Bender, P., and Alberty, R. A. (1970). *Experimental Physical Chemistry*, 7th ed. McGraw-Hill, New York.

da Silva, J. A. L., Goncalves, M. P., and Rao, M. A. (1993). Viscoelastic behavior of mixtures of locust bean gum and pectin dispersions. *J. Food Eng.* 18:211–228.

Dautzenberg, H., Jaeger, W., Kotz, J., Philipp, B., Seidel, C., and Stscherbina, D. (1994). *Polyelectrolytes: Formation, Characterization and Application*. Hanser Verlag, Munich.

Dea, I. C. M. (1987). The role of structural modification in controlling polysaccharide functionality. In *Industrial Polysaccharides: Genetic Engineering, Structure/Property Relations and Applications*, Yalpini, I. (Ed.), pp. 207–216. Elsevier, Amsterdam/New York.

Dea, I. C. M. (1989). Industrial polysaccharides. *Pure Appl. Chem.* 61:1315–1322.

Dea, I. C. M., and Morris, E. R. (1977). Synergistic xanthan gels. In *Extracellular Microbial Polysaccharides*, Sandford, P. A., and Laskin, A. (Eds.), ACS Symposium Series 45, pp. 174–182. Am. Chem. Soc., Washington, DC.

Dea, I. C. M., McKinnon, A. A., and Rees, D. A. (1972). Tertiary and quaternary structure in aqueous polysaccharide systems which model cell wall cohesion: reversible changes in conformation and association of agarose, carrageenan and galactomannans. *J. Mol. Biol.* 68:153–172.

Debye, P. (1944). Light scattering in solutions. *J. Appl. Phys.* 15:338–342.

Debye, P. (1947). Molecular-weight determination by light scattering. *J. Phys. Colloid Chem.* 51:18–32.

Deckers, H. A., Olieman, C., Rombouts, F. M., and Pilnik, W. (1986). Calibration and application of high-performance size-exclusion columns for molecular-weight distribution of pectins. *Carbohydr. Res.* 6:361–378.

de Gennes, P. G. (1979). *Scaling Concepts in Polymer Physics*, Cornell Univ. Press, Ithaca, NY.

Dekker, R. F. H. (1979). The hemicellulase group of enzymes. In *Polysaccharides in Food*, Blanshard, J. V. M., and Mitchell, J. R. (Eds.), pp. 93–96. Butterworth, London.

Dentini, M., Crescenzi, V., and Matricardi, P. (1991). A novel procedure for determining the average charge density of pectin chains. *Food Hydrocoll.* 5:307–312.

Deslandes, Y., Marchessault, R. H., and Sarko, A. (1980). Triple-helical structure of (1,3)-β-D-glucan. *Macromolecules* 13:1466–1471.

Desmarais, A. J., and Wint, R. F. (1993). Hydroxyalkyl and ethyl ethers of cellulose. In *Industrial Gums*, 3rd ed., Whistler, R. L., and BeMiller, J. N. (Eds.), pp. 505–535. Academic Press, New York.

De Vries, J. A., Rombouts, F. M., Voragen, A. G. J., and Pilnik, W. (1982). Enzymic degradation of apple pectins. *Carbohydr. Polym.* 2:25–33.

De Vries, J. A., Voragen, A. G. J., Rombouts, F. M., and Pilnik, W. (1981). Extraction and purification of pectins from alcohol-insoluble solids from ripe and unripe apples. *Carbohydr. Polym.* 1:117–127.

de Willigen, A. H. A. (1964). Potato starch. *Methods in Carbohydrate Chemistry*, Vol. IV, pp. 9–13. Academic Press, New York.

Dickinson, E. (1988). The role of hydrocolloids in stabilizing particulate dispersions and emulsions. In *Gums and Stabilizers for the Food Industry*, Phillips, G. O., Williams, P. A., and Wedlock, D. J. (Eds.), Vol. 4, pp. 249–263. IRL Press, Oxford.

Dickinson, E. (1992). *An Introduction to Food Colloids*. Oxford Univ. Press, London.

Dickinson, E. (1993). Protein–polysaccharide interactions in food colloids. In *Food Colloids: Stability and Mechanical Properties*, Dickinson, E., and Walstra, P. (Eds.), pp. 77–93. Royal Chem. Soc., London.

Dickinson, E., and Euston, S. R. (1991a). Computer simulation of macromolecular adsorption. In *Food Polymers, Gels and Colloids*, Dickinson, E. (Ed.), pp. 557–563. Royal Chem. Soc., London.

Dickinson, E., and Euston, S. R. (1991b). Stability of food emulsions containing both protein and polysaccharide. In *Food Polymers, Gels and Colloids*, Dickinson, E. (Ed.), pp. 132–145. Royal Chem. Soc., London.

Dickinson, E., and Galazka, V. B. (1991). Bridging flocculation in emulsions made with a mixture of protein + polysaccharide. In *Food Polymers, Gels and Colloids*, Dickinson, E. (Ed.), pp. 494–497. Royal Chem. Soc., London.

Dickinson, E., and Stainsby, G. (1982). *Colloids in Food*, p. 468. Appl. Sci., Braking, Essex.

Dickinson, E., Elverson, D. J., and Murray, B. S. (1989). On the film-forming and emulsion-stabilizing properties of gum arabic: dilution and flocculation aspects. *Food Hydrocoll.* 3:101–114.

Dickinson, E., Galazka, V. B., and Anderson, D. M. W. (1991a). Emulsifying behavior of gum arabic. Part I. Effect of the nature of the oil phase on the emulsion droplet-size distribution. *Carbohydr. Polym.* 14:373–383; Part II. Effect of the gum molecular weight on the emulsion droplet-size distribution. *Carbohydr. Polym.* 14:385–392.

Dickinson, E., Galazka, V. B., and Anderson, D. M. W. (1991b). Effect of molecular weight on the emulsifying behavior of gum arabic. In *Food Polymers, Gels and Colloids*, Dickinson, E. (Ed.), pp. 490–493. Royal Chem. Soc., London.

Dickmann, R. S., Chism, G. W., Renoll, M. W., and Hansen, P. M. T. (1989). Detection of food gums on polyamide strips using horseradish peroxidase–benzidine staining and laser-beam densitometry. *Food Hydrocoll.* 3:33–40.

Dische, Z. (1962). Color reactions of carbohydrates. In *Methods in Carbohydrate Chemistry*, Whistler, R. L., Wolfrom, M. L., BeMiller, J. N., and Shafizadeh, F. (Eds.), Vol. I, pp. 475–514. Academic Press, New York.

Dische, Z., and Shettles, L. B. (1948). A specific color reaction of methylpentoses and a spectrophotometric micromethod for their determination. *J. Biol. Chem.* 175:595–603.

Doesburg, J. J. (1965). Pectic Substances in Fresh and Preserved Fruits and Vegetables. Communication No. 25, I.B.V.T., Wageningen, The Netherlands.

Doi, M., and Edwards, S. F. (1986). *The Theory of Polymer Dynamics*. Clarendon, Oxford.

Donnelly, B. J., Voight, J. E., and Scallet, B. L. (1980). Reactions of oligosaccharides. V. Pyrolysis–gas chromatography. *Cereal Chem.* 57:388–389.

Doty, P., Wada, A., Yang, J. T., and Blout, E. R. (1957). Polypeptides VII. Molecular configurations of poly-L-glutamic acid in water–dioxane solution. *J. Polym. Sci.* 23:851–861.

Doublier, J.-L., and Llamas, G. (1991). Flow and viscoelastic properties of mixed xanthan gum + galactomannan systems. In *Food Polymers, Gels and Colloids*, Dickinson, E. (Ed.), pp. 349–356. Royal Chem. Soc., London.

Doublier, J.-L., and Llamas, G. (1993). A rheological description of amylose–amylopectin mixtures. In *Food Colloids and Polymers: Stability and Mechanical Properties*, Dickinson, E., and Walstra, P. (Eds.), pp. 138–146. Royal Chem. Soc., London.

Dow Chemical Co. (1990). A Food Technologist's Guide to Methocel Premium Food Gums. Form No. 192-1037-190X-AMS, Dow Chemical Co., Midland, MI.

DuPont Co. (n.d.). Sodium CMC: Carboxylmethylcellulose: Basic Properties. E.I. Dupont & Co., Wilmington, DE.

Duxbury, D. D. (1993). Fat reduction without adding fat replacers. *Food Processing USA* 54:68, 70.

Dziezak, J. D. (1991). A focus on gums. *Food Technol.* 45:116, 118–120, 122–124, 126, 128, 130, 132.

Edwards, S. F. (1966). The theory of polymer solutions at intermediate concentration. *Proc. Phys. Soc., London* 85:613–624.

Eisenberg, A., and King, M. (1977). *Ion-Containing Polymers: Physical Properties and Structure*. Academic Press, New York.

Elbirli, B., and Shaw, M. T. (1978). Time constants from shear viscosity data. *J. Rheol.* 22: 561–570.

Elfak, A. M., Pass, G., Phillips, G. O., and Morley, R. G. (1977). The viscosity of dilute solutions of guar and locust bean gum with and without added sugars. *J. Sci. Food. Agric.* 28:895–899.

Elias, H.-G. (1979). *Macromolecules*, Vol. I. Plenum, New York.

Eliasson, A.-C. (1992). A calorimetric investigation of the influence of sucrose on the gelatinization of starch. *Carbohydr. Polym.* 18:131–138.

Elworthy, P. H., and George, T. M. (1963). The molecular properties of ghatti gum: a naturally occurring polyelectrolyte. *J. Pharm. and Pharmacol.* 15:781–793.

Engster, M., and Abraham, R. (1976). Cecal response to different molecular weights and types of carrageenan in the guinea pig. *Toxicol. Appl. Pharmacol.* 38:265–282.

Everett, D. H. (1988). *Basic Principles of Colloid Science*, p. 153. Royal Chem. Soc., London.

Eyring, H. (1936). Viscosity, plasticity, and diffusion as examples of absolute reaction rates. *J. Chem. Phys.* 4:283-291.

Fagerson, I. S. (1969). Thermal degradation of carbohydrates. *J. Agric. Food Chem.* 17:747–750.

Fang, P., McGinnis, G. D., and Wilson, W. W. (1981). Thermal fragmentation for the identification of pyrolytic products from bark polysaccharides. *Anal. Chem.* 53:2172–2174.

Fasihuddin, B. A., Wedlock, D. J., Omar, S., and Phillips, G. O. (1988). Solution properties of sodium alginate. In *Gums and Stabilizers for the Food Industry*, Phillips, G. O., Williams, P. A., and Wedlock, D. J. (Eds.), Vol. 4, pp. 89–96. IRL Press, Oxford.

Feddersen, R. L., and Thorp, S. N. (1993). Sodium carboxymethylcellulose. In *Industrial Gums*, Whistler, R. L., and BeMiller, J. N. (Eds.), pp. 537–578. Academic Press, San Diego.

Ferry, J. D. (1980). *Viscoelastic Properties of Polymers*, 3rd ed. Wiley, New York.

Finkelmann, H., and Jahns, E. (1989). Association and liquid crystalline phases of polymers in solution. In *Polymer Association Structures*, El-Nokaly, M. A. (Ed.), ACS Symposium Series 384, pp. 1–20. Am. Chem. Soc., Washington, DC.

Fishman, M. L., El-Ataway, Y. S., Sondey, S. M., Gillespie, D. T., and Hicks, K. B. (1991). Component and global average radii of gyration of pectins from various sources. *Carbohydr. Polym.* 15:89–104.

Fishman, M. L., Pepper, L., and Pfeffer, P. E. (1986). Dilute solution properties of pectin. In *Water-Soluble Polymers*, Glass, J. E., Jr., Advances in Chemistry 213. Am. Chem. Soc., Washington, DC.

Fiszman, S. M., Costell, E., and Duran, L. (1986). Effects of the addition of sucrose and cellulose on the compression behavior of carrageenan gels. *Food Hydrocoll.* 1:113–120.

Flexel, Inc. (1989). *PUT Technical Data Sheet*. Flexel, Inc., Atlanta, GA.

Flory, P. J. (1953). *Principles of Polymer Chemistry*. Cornell Univ. Press, Ithaca, NY.

FMC, (1989). Nutricol Konjac Flour. Bulletin K-1, FMC Corporation, Philadelphia, PA.

FMC (1991). Marine colloids products balance taste, texture and mouthfeel in reduced fat foods. *The Carrageenan People (FMC Corp.)* 8.

FMC (1993). *Avicel Cellulose Gel (Microcrystalline Cellulose): General Technology*. FMC Corporation, Philadelphia, PA; *Novagel Cellulose Gel*, FMC Corporation, Philadelphia, PA.

Folkes, D. J. (1980). A gas chromatographic method for the determination of pentosans. *J. Sci. Food Agric.* 31:1011–1016.

Food Product Development (1979). Basic guidelines for food gum selection. *Food Product Dev.* February, 21, 24, 26, 28.

Food Technology (1990). Gellan gum receives FDA approval. *Food Technol.* 44:88.

Food Technology (1992). Starches and gums. *Food Technol.* 46:174, 176, 178.

Frank, H. P., and Mark, H. F. (1955). Report on molecular-weight measurements of standard polystyrene samples. II. International Union of Pure and Applied Chemistry. *J. Polym. Sci.* 17:1–20.

Freudenberg, K., Schaaf, E., Dumbert, G., and Ploetz, T. (1939). *Naturwissenschaften* 27:850. Cited by Langlois, D. P., and Wagoner, J. A. (1967). Production and use of amylose. In *Starch: Chemistry and Technology*, Whistler, R. L., and Paschall, E. F. (Eds.), p. 463. Academic Press, New York.

Furusawa, K., Kimura, Y., and Tagawa, T. (1984). In *Polymer Adsorption and Dispersion Stability*, Goddard, E. D., and Vincent, B. (Eds.), ACS Symposium Series 240. Am. Chem. Soc., Washington, DC.

Galloway, G. I., Biliaderis, C. G., and Stanley, D. W. (1989). Properties and structure of amylose–glyceryl monostearate complexes formed in solution or on extrusion of wheat flour. *J. Food Sci.* 54:950–957.

Ganz, A. J. (1974). How cellulose gum reacts with proteins. *Food Eng.* 46:67–69.

Garmon, R. G. (1975). End group determinations. In *Polymer Molecular Weights*, Slade, P. E., Jr. (Ed.), Vol. I, p. 34. Dekker, New York.

Geddes, A. L. (1949). Determination of diffusivity. In *Physical Methods of Organic Chemistry*, 2nd ed., Weissberger, A. (Ed.), p. 617. Interscience, New York.

Gekko, K., Mugishima, H., and Koga, S. (1987). Effects of sugars and polyols on the sol-gel transition of κ-carrageenan: calorimetric study. *Int. J. Biol. Macromol.* 9:146–152.

Gelman, R. A. (1982). Characterization of carboxymethylcellulose: distribution of substituent groups along the chain. *J. Appl. Polym. Sci.* 27:2957–2964.

Giddings, J. C., Myers, M. N., Caldwell, K. D., and Fisher, S. R. (1980). Analysis of biological macromolecules and particles by field-flow fractionation. In *Methods of Biochemical Analysis*, Glick, D. (Ed.), Vol. 26, pp. 79–136. Wiley, New York.

Gidley, M. J. (1988). Conformational studies of α-(1,4) glucans in solid and solution states by NMR spectroscopy. In *Gums and Stabilizers for the Food Industry*, Phillips, G. O., Williams, P. A., and Wedlock, D. J. (Eds.), Vol. 6, pp. 71–80. IRL Press, Oxford.

Gidley, M. J. (1989). Molecular mechanisms underlying amylose aggregation and and gelation. *Macromolecules* 22:351–358.

Gidley, M. J. (1992). Nuclear magnetic resonance analysis of cereal carbohydrates. In *Developments in Carbohydrate Chemistry*, Alexander, R. J., and Zobel, H. F. (Eds.), pp. 163–191. Amer. Assoc. Cereal Chem., St. Paul, MN.

Gilliland, E. R., and Gutoff, E. B. (1960). Rubber-filler interactions: solution adsorption studies. *J. Appl. Polym. Sci.* 3:26–42.

Glass, J. E. (1986). Structural features promoting water solubility in carbohydrate polymers. In *Water-Soluble Polymers*, Glass, J. E. (Ed.), Advances in Chemistry 213, pp. 3–27. Am. Chem. Soc., Washington, DC.

Glasstone, S., and Lewis, D. (1960). *Elements of Physical Chemistry*. Van Nostrand, New York.

Glicksman, M. (1969). *Gum Technology in the Food Industry*. Academic Press, New York.

Glicksman, M. (1982). *Food Hydrocolloids*, Vol. I. CRC Press, Boca Raton, FL.

Glicksman, M. (1983). *Food Hydrocolloids*, Vols. II and III. CRC Press, Boca Raton, FL.

Glicksman, M. (1991). Hydrocolloids and the search for the oily grail. *Food Technol.* 45:94, 96–101.

Gough, B. M., and Pybus, J. N. (1971). Effect on the gelatinization temperature of wheat starch granules of prolonged treatment with water at 50°C. *Die Stärke* 23:210–212.

Gould, E. S. (1962). *Inorganic Reactions and Structure*. Holt, Rinehart & Winston, New York.

Gould, J. M., Jasberg, B. K., and Dexter, L. (1990). Effects of alkaline hydrogen peroxide-treated fiber ingredients on mixograph properties of wheat flour dough. *Lebensm. Wiss. Technol.* 23:358–360.

Graham, H. D. (1971). Determination of carboxymethylcellulose in food products. *J. Food Sci.* 36:1052–1055.

Graham, H. D. (1977). Analytical methods for major plant hydrocolloids. In *Food Colloids*, Graham, H. D. (Ed.), pp. 540–579. AVI Press, New York.

Graham, H. D. (1990). Semi-micro method for the quantitative determination of gellan gum in food products. *Food Hydrocoll.* 3:435–445.

Grant, G. T., Morris, E. R., Rees, D. A., Smith, P. J. C., and Thom, D. (1973). Biological interactions between polysaccharides and divalent cations: the egg-box model. *Fed. Exper. Biol. Sci. Lett.* 32:195.

Grasdalen, H., and Painter, T. (1980). N.M.R. studies of composition and sequence in legume-seed galactomannans. *Carbohydr. Res.* 81:59–66.

Green, J. W. (1963). Wood cellulose. *Methods in Carbohydrate Chemistry*, Vol. III, pp. 9–21. Academic Press, New York; Drying and reactivity of cellulose. *Methods in Carbohydrate Chemistry*, Vol. III, pp. 101–103. Academic Press, New York.

Greenish, H. G. (1923). *Foods and Drugs*, 3rd ed. Churchill, London.

Greenwood, C. T. (1967). The thermal degradation of starch. *Adv. Carbohydr. Chem.* 22:498–500.

Greenwood, C. T., Knox, J. H., and Milne, E. (1961). Analysis of the thermal decomposition products of carbohydrates by gas-chromatography. *Chem. Ind. (London)* 1878–1879.

Griffiths, A. J., and Kennedy, J. F. (1988). Biotechnology of polysaccharides. In *Carbohydrate Chemistry*, Kennedy, J. F. (Ed.), p. 625. Clarendon, Oxford.

Grinberg, V. Ya., and Tolstogusov, V. B. (1972). Thermodynamic compatibility of gelatin with some D-glucans in aqueous media. *Carbohydr. Res.* 25:313–321.

Grover, J. A. (1993). Methylcellulose and derivatives. In *Industrial Gums*, 3rd ed., Whistler, R. L., and BeMiller, J. N. (Eds.), pp. 475–504. Academic Press, San Diego.

Grubisic, Z., Rempp, P., and Benoit, H. (1967). A universal calibration for gel permeation chromatography. *Polym. Lett.* 5:753–759.

Guiseley, K. B., Stanley, N. F., and Whitehouse, P. A. (1980). Carrageenan. In *Handbook of Water-Soluble Gums and Resins*, Davidson, R. L. (Ed.), pp. 5-1–5-30. McGraw-Hill, New York.

Guo, J.-X., and Gray, D. G. (1991). Spectral behavior of dye molecules oriented by chiral nematic (acetyl)(ethyl)cellulose solutions. In *Fourth Chemistry Congress of North America*, Polym. Abstract 331. Am. Chem. Soc., Washington, DC.

Guo, J.-X., and Gray, D. G. (1994). Lyotropic cellulose liquid crystals. In *Cellulosic Polymers: Blends and Composites*, Gilbert, R. D. (Ed.), p. 27. Hanser/Verlag, Munich.

Ha, Y. W., and Thomas, R. L. (1988). Simultaneous determination of neutral sugars and uronic acids in hydrocolloids. *J. Food Sci.* 53:574–577.

Haas, R., and Kulicke, W. M. (1984). Flow behavior of dilute polyacrylamide solutions through porous media. 2. Indirect determination of extremely high molecular weights and some aspects of viscosity decrease over long time intervals. *Ind. Eng. Chem. Fund* 23:316–319.

Hammett, L. P. (1952). *Introduction to the Study of Physical Chemistry*. McGraw-Hill, New York.

Hannay, N. B. (1967). *Solid-State Chemistry*, p. 104. Prentice-Hall, Englewood Cliffs, NJ.

Hansen, P. M. T., and Whitney, R. M. (1960). A quantitative test for carrageenan ester sulfate in milk products. *J. Dairy Sci.* 43:175–186.

Harada, A., Li, J., and Kamachi, M. (1993). Synthesis of a tubular polymer from threaded cyclodextrins. *Nature (London)* 364:516–518.

Harada, T., Kanzawa, Y., Kanenaga, K., Koreeda, A., and Harada, A. (1991). Electron microscopic studies on the ultrastructure of curdlan and other polysaccharides in gels used in foods. *Food Struct.* 10:1.

Harada, T., Terasaki, M., and Harada, A. (1993). Curdlan. In *Industrial Gums*, Whistler, R. L., and BeMiller, J. N. (Eds.), pp. 427–445. Academic Press, New York.

Harding, S. E., Berth, G., Ball, A., and Mitchell, J. R. (1990). Hydrodynamic evidence for an extended conformation for citrus pectins in dilute solution. In *Gums and Stabilizers for the Food Industry*, Phillips, G. O., Wedlock, D. J., and Williams, P. A. (Eds.), Vol. 5, pp. 267–271. IRL Press, Oxford.

Harding, S. E., Berth, G., Ball, A., Mitchell, J. R., and Garcia de la Torre, J. (1991b). The molecular weight distribution and conformation of citrus pectins in solution studied by hydrodynamics. *Carbohydr. Polym.* 16:1–15.

Harding, S. E., Vårum, K. M., Stokke, Bjørn, T., and Smidsrød, O. (1991a). Molecular weight determination of polysaccharides. *Advances in Carbohydrate Analysis*, White, C. A. (Ed.), Vol. 1. JAI Press, Greenwich, CT.

Harris, P., Morrison, A., and Dacombe, C. (1995). A practical approach to polysaccharide analysis. In *Food Polysaccharides and Their Applications*, Stephen, A. M. (Ed.), pp. 577–606. Dekker, New York.

Hart, R. J., Lynch, G., Dea, I. C. M., and Morris, E. R. (1992). Manipulation and control of mixed polysaccharide milk protein gels. In *Gums and Stabilizers for the Food Industry*, Phillips, G. O., Williams, P. A., and Wedlock, D. J. (Eds.), Vol. 6, pp. 173–179. IRL Press, Oxford.

Hatakeyama, T., Nakamura, K., Yoshida, H., and Hatakeyama, H. (1989). Mesomorphic properties of highly concentrated aqueous solutions of polyelectrolytes from saccharides. *Food Hydrocoll.* 3:301–311.

Hayashi, A., and Kanzaki, T. (1987). Swelling of agarose gel and its related changes. *Food Hydrocoll.* 1:317–325.

Hebeda, R. E., and Teague, W. M. (1994). Starch hydrolyzing enzymes. In *Developments in Carbohydrate Chemistry*, Alexander, R. J., and Zobel, H. F. (Eds.), pp. 65–85. Amer. Assoc. Cereal Chem., St. Paul, MN.

Hedges, A. R., Shieh, W. J., and Sikorski, C. T. (1993). Use of cyclodextrins for encapsulation in the use and treatment of food products. 206th National Meeting, Am. Chem. Soc., Div. Agric. Food Chem., Chicago, IL, Abstract 95.

Hegenbart, S. (1989). New ideas gel with starches and gums. *Prepared Foods* 158:105, 106, 108.

Heller, W. (1966). Effects of macromolecular compounds in disperse systems. *Pure Appl. Chem.* 12:249–274.

Hellerqvist, C. G., and Sweetman, B. J. (1990). Mass spectrometry of carbohydrates. In *Methods of Biochemical Analysis: Biomedical Applications of Mass Spectrometry*, Suelter, C. H., and Watson, J. T. (Eds.), Vol. 34, pp. 91–143. Wiley, New York.

Henderson, A. (1988). Cellulose ethers—the role of thermal gelation. In *Gums and Stabilizers for the Food Industry*, Phillips, G. O., Williams, P. A., and Wedlock, D. J. (Eds.), Vol. 4, pp. 265–275. IRL Press, Oxford.

Henderson, A. (1989). Properties of MC and HPMC cellulose derivative gums. *Food Bev. Technol. International, USA*, pp. 163–167.

Hercules, Inc. (1971). *Klucel Hydroxypropyl Cellulose*. Hercules, Inc., Wilmington, DE.

Hercules, Inc. (1980). *Cellulose Gum*. Hercules, Inc., Wilmington, DE.

Hercules, Inc. (1985). *Genu Handbook for the Fruit Processing Industry*, p. 14-1. Hercules, Inc., Middletown, New York.

Hersom, A. C., and Hulland, E. D. (1981). *Canned Foods*, 7th ed. Chem. Publ. Co., New York.

Hestrin, S. (1963). Bacterial cellulose. *Methods in Carbohydrate Chemistry*, Vol. III, pp. 4–9. Academic Press, New York.

Heyn, A. N. J. (1966). The microcrystalline structure of cellulose in cell walls of cotton, ramie, and jute fibers as revealed by negative staining of sections. *J. Cell. Biol.* 29:181–197.

Heyns, K., Stute, R., and Paulson, H. (1966). Braunungsreaktionen und fragmentierungen von kohlenhydraten. *Cabohydr. Res.* 2:132–149.

Heyraud, A., and Rinaudo, M. (1991). Use of multidetection for chromatographic characterization of dextrins and starch. In *Biotechnology of Amylodextrin Oligosaccharides*, Friedman, R. B. (Ed.), ACS Symposium Series 458, pp. 171–188. Amer. Chem. Soc., Washington, DC.

Hicks, K. B., Lim, P. C., and Haas, M. J. (1985). Analysis of uronic and aldonic acids, their lactones, and related compounds by high-performance liquid chromatography on cation-exchange resins. *J. Chromatogr.* 319:159–171.

Hickson, T. G. L., and Polson, A. (1968). Some physical characteristics of the agarose molecule. *Biochem. Biophys. Acta* 165:43–58.

Hiemenz, P. C. (1986). *Principles of Colloid and Surface Chemistry*, 2nd ed. Dekker, New York.

Hinton, C. L. (1950). The setting temperature of pectin jellies. *J. Sci. Food Agric.* 1:300.

Hodge, J. E. (1953). Chemistry of browning reactions in model systems. *J. Agric. Food Chem.* 1:928–943.

Hoefler, A. C. (1991). Other pectin food products. In *The Chemistry and Technology of Pectin*, Walter, R. H. (Ed.), pp. 51–66. Academic Press, New York.

Holmes, F. H., and Shaw, C. J. G. (1961). The pyrolysis of cellulose and the action of flame-retardants. I. Significance and analysis of the tar. *J. Appl. Chem.* 11:210–216.

Holmes, H. N. (1922). *Laboratory Manual of Colloid Chemistry*, p. 22. Wiley, New York.

Horton, D. (1965). Pyrolysis of starch. In *Starch Chemistry and Technology*, Whistler, R. L., and Paschall, E. F. (Eds.), Vol. I, pp. 421–437. Academic Press, New York.

Horton, J. C., and Donald, A. M. (1991). Gelation in a synthetic polypeptide system. In *Food Polymers, Gels and Colloids*, Dickinson, E. (Ed.), pp. 508–512. Royal Chem. Soc., London.

Horton, J. C., Harding, S. E., Mitchell, J. R., and Morton-Holmes, D. F. (1991). Thermodynamic non-ideality of dilute solutions of sodium alginate studied by sedimentation equilibrium ultracentrifugation. *Food Hydrocoll.* 5:125–127.

Horton, S. D., Lauer, G. N., and White, J. S. (1990). Predicting gelatinization temperatures of starch/sweetener systems for cake formulation by differential scanning calorimetry. II. Evaluation and application of a model. *Cereal Foods World* 35:734–737, 739.

Hoseney, R. C. (1986). *Principles of Cereal Science and Technology*. Amer. Assoc. Cereal Chem., St. Paul, MN.

Houminer, Y. (1973). Thermal degradation of carbohydrates. In *Molecular Structure and Function of Food Carbohydrates*, Birch, G. G., and Green, L. F. (Eds.), pp. 133–153. Appl. Sci., Braking, Essex.

Hourdet, D., and Muller, G. (1991). Solution properties of pectin polysaccharides II. Conformation and molecular size of high galacturonic acid content isolated pectin chains. *Carbohydr. Polym.* 16:113–135.

Howling, D. (1980). The influence of the structure of starch on its rheological properties. *Food Chem.* 6:51–61.

Huggins, M. L. (1942). The viscosity of dilute solutions of long-chain molecules. IV. Dependence on concentration. *J. Am. Chem. Soc.* 64:2716–2718.

Hui, P. A., and Neukom, H. (1964). Some properties of galactomannans. *Tech. Assoc. Pulp and Paper Indus.* 47:39–42.

Hunkeler, D., Wu, X. Y., and Hamielec, A. E. (1992). In *Polyelectrolyte Gels*, Harland, R. S., and Prud'homme, R. K. (Eds.), ACS Symposium Series 480, pp. 53–79. Amer. Chem. Soc., Washington, DC.

Hwang, J., and Kokini, J. (1991). Structure and rheological function of side branches of carbohydrate polymers. *J. Texture Studies* 22:123–167.

IFT (1959). Pectin standardization. Final Report of the Institute of Food Technologists Committee. *Food Technol.* 13:496–500.

Ikan, R. (1991). *Natural Products*, 2nd ed., p. 76. Academic Press, New York.

Ilmain, F., Tanaka, T., and Kokufuta, E. (1991). Volume transition in a gel driven by hydrogen bonding. *Nature (London)* 349:400–401.

Imeson, A. P., Ledward, D. A., and Mitchell, J. R. (1977). On the nature of the interaction between some anionic polysaccharides and proteins. *J. Sci. Food Agric.* 28:661–668.

Ingle, T. R., and Whistler, R. L. (1964). End-group analysis by methylation. *Methods In Carbohydrate Chemistry*, Vol. IV, pp. 83–86. Academic Press, New York.

Inglett, G. E., and Grisamore, S. B. (1991). Maltodextrin fat substitute lowers cholesterol. *Food Technol.* 45:104.

Irwin, W. J. (1979). Analytical pyrolysis—an overview. *J. Anal. Appl. Pyrolysis* 1:3–25, 89–122.

Irwin, W. J. (1982). *Analytical Pyrolysis*, pp. 339–352. Dekker, New York.

Ishii, S., Kiho, K., Sugiyama, S., and Sugimoto, H. (1979). Low-methoxyl pectin prepared by pectinesterase from *Aspergillus japonicus*. *J. Food Sci.* 44:611–614.

Isogai, A. (1994). Allomorphs of cellulose and other polysaccharides. In *Cellulose Polymers: Blends and Composites*, Gilbert, R. D. (Ed.), p. 21. Hanser Verlag, Munich.

Israelachvili, J. N. (1992). *Intermolecular and Surface Forces*, 2nd ed. Academic Press, New York.

Jacon, S. A., Rao, M. A., Cooley, H. J., and Walter, R. H. (1993). The isolation and characterization of a water extract of konjac flour gum. *Carbohydr. Polym.* 20:35–41.

Jansson, P.-E., and Lindberg, B. (1983). Structural studies of gellan gum, an extracellular polysaccharide elaborated by *Pseudomonas elodea*. *Carbohydr. Res.* 124:135–139.

Javeri, H. Toledo, R., and Wicker, L. (1991). Vacuum infusion of citrus pectinmethylesterase and calcium effects on firmness of peaches. *J. Food Sci.* 56:739–742.

Jayme, G., and Lang, F. (1963). Cellulose solvents. In *Methods in Carbohydrate Chemistry*, Vol. III, pp. 75–83. Academic Press, New York.

Jeffrey, G. A., and Lewis, L. (1978). Cooperative aspects of hydrogen bonding in carbohydrates. *Carbohydr. Res.* 60:179–182.

Jeltema, M. A., Zabic, M. E., and Thiel, L. J. (1983). Prediction of cookie quality from dietary fiber components. *Cereal Chem.* 60:227–230.

Jirgensons, B. (1946). Dependence of solvation on particle shape. *J. Polym. Sci.* 1:475–483.

Jirgensons, B., and Straumanis, M. E. (1962). *A Short Textbook of Colloid Chemistry*, pp. 159, 377. Macmillan, New York.

Johnson, D. C., Nicholson, M. D., and Haigh, F. C. (1975). Dimethylsulfoxide/Paraformaldehyde: A Nondegrading Solvent for Cellulose. Technical Paper Series No. 5, Inst. Paper Chem., Appleton, WI.

Johnson, W. C. (1987). The circular dichroism of carbohydrates. *Adv. Carbohydr. Chem. Biochem.* 45:85–92.

Johnston, F. B. (1956). Isolation of plant starches. *Nature (London)* 178:370.

Jones, J. K. N., and Smith, F. (1949). Plant gums and mucilages. *Adv. Carbohydr. Chem.* 4:244.

Jones, S. A. (1996). Issues in fat replacement. In *Handbook of Fat Replacers*, Roller, S., and Jones, S. A. (Eds.), pp. 3–26. CRC Press, Boca Raton, FL; Physical, chemical, and sensory aspects of fat replacement. In *Handbook of Fat Replacers*, Roller, S., and Jones, S. A. (Eds.), pp. 59–86. CRC Press, Boca Raton, FL.

Jordan, R. C., and Brant, D. A. (1978). An investigation of pectin and pectic acid in dilute aqueous solution. *Biopolymers* 17:2885–2895.

Kaelble, D. H. (1971). *Physical Chemistry of Adhesion*, pp. 327–330. Wiley-Interscience, New York.

Kalichevsky, M. T., and Ring, S. G. (1987). Incompatibility of amylose and amylopectin in aqueous solution. *Carbohydr. Res.* 162:323–328.

Kalichevsky, M. T., Orford, P. D., and Ring, S. G. (1986). The incompatibility of concentrated aqueous solutions of dextran and amylose and its effect on amylose gelation. *Carbohydr. Polym.* 6:145–154.

Kanzaki, G., and Berger, E. Y. (1959). Colorimetric determination of methylcellulose with diphenylamine. *Anal. Chem.* 31:1383–1385.

Kanzawa, Y., Koreeda, A., Harada, A., and Harada, T. (1989). Electron microscopy of the gel-forming ability of polysaccharide food additives. *Agric. Biol. Chem.* 53:979–986.

Kasai, N., and Harada, T. (1979). Ultrastructure of curdlan. *Chem. Abstr.* 178(1).

Kato, K., and Matsuda, K. (1973). Isolation of oligosaccharides corresponding to the branching point of konjac mannan. *Agric. Biol. Chem.* 37:2045–2051.

Keijbets, M. J. H. (1974). Pectic Substances in the Cell Wall and the Intercellular Cohesion of Potato Tuber Tissue during Cooking. Ph.D. thesis, Centre for Agricultural Publishing and Documentation, Wageningen, The Netherlands.

Kelco (1976). *Xanthan Gum*, 2nd ed. Kelco, Division of Merck, Rahway, NJ.

Kelco (1986). Structured Foods with the Algin/Calcium Reaction. Technical Bulletin, F-83, Kelco, San Diego, CA.

Kennedy, J. F., Melo, E. H. M., Crescenzi, V., Dentini, M., and Matricardi, P. (1992). A rapid quantitative determination of pectin and carboxymethyl cellulose in solution using poly(hexamethylenebiguanidinium) chloride. *Carbohydr. Polym.* 17:199–203.

Kerner, P., and Ward, F. M. (1992). Blend of hydrocolloids mimics fats and oils. In *Food Technol.* 46:156.

Kerr, R. W. (1950). *Chemistry and Industry of Starch*, 2nd rev. ed. Academic Press, New York.

Kester, J. J., and Fennema, O. R. (1986). Edible films and coatings: a review. *Food Technol.* 40:47–59.

Khoury, F., and Passaglia, E. (1976). The morphology of crystalline synthetic polymers. *Treatise on Solid State Chemistry*, Hannay, N. B. (Ed.), Vol. 3, pp. 335–496. Plenum, New York.

Kim, Y. J., and Kim, C. S. (1988). Effect of apple hemicellulose on calcium pectate gel formation. *J. Korean Soc. Food Nutr.* 17:13–17 (*Food Sci. Technol. Abstr.* 12T0029, 1989).

Kimura, H., Moritaka, S., and Masaru, M. (1973). Polysaccharide 13140: a new thermo-gelable polysaccharide. *J. Food Sci.* 38:668–670.

King, K. (1994). Changes in the functional properties and molecular weight of sodium alginate following γ irradiation. *Food Hydrocoll.* 8:83–96.

King, K., and Gray, R. (1993). The effect of gamma irradiation on guar gum, locust bean gum, gum tragacanth and gum karaya. *Food Hydrocoll.* 6:559–569.

Kintner, P. I., and Van Buren, J. P. (1982). Carbohydrate interference and its correction in pectin analysis using the *m*-hydroxydiphenyl method. *J. Food Sci.* 47:756–759, 764.

Kipling, J. J. (1965). *Adsorption from Solutions of Non-Electrolytes*, pp. 288–298. Academic Press, New York.

Kishida, N., Okimasu, S., and Kamata, T. (1978). Molecular weight and intrinsic viscosity of konjac gluco-mannan. *Agric. Biol. Chem.* 42:1645–1650.

Klemann, L. P., and Finley, J. W. (1989). Low calorie fat mimetics comprising carboxyl/carboxylate esters. Patent application (*Food Sci. Technol. Abstr.* 21, No. 11V110).

Klockeman, D. M., Pressey, R., and Jen, J. J. (1991). Characterization of Cell Wall Polysaccharides of Jicama (*Pachyrrhizus erosus*) and Chinese Water Chestnut (*Eleocharis dulicis*), Abstract 75, Agric. & Food Chem. Div., Am. Chem. Soc., Washington, DC.

Klyosov, A. A. (1991). Trends in biochemistry and enzymology in cellulose degradation. *Biochemistry* 29:10577–10585.

Knight, A. R. (1970). *Introductory Physical Chemistry*, pp. 46–54. Prentice-Hall, Englewoods Cliff, NJ.

Knirel, Y., Vinogradov, E. V., and Mort, A. J. (1989). Application of anhydrous hydrogen fluoride for structural analysis of polysaccharides. *Adv. Carbohydr. Chem. Biochem.* 47:202.

Knorr, D., Wampler, T. P., and Teutonico, R. A. (1985). Formation of pyrazines by chitin pyrolysis. *J. Food Sci.* 50:1762–1763.

Kobayashi, S., Schwartz, S. J., and Linebeck, D. R. (1985). Rapid analysis of starch, amylose and amylopectin by high-performance size-exclusion chromatography. *J. Chromatogr.* 319: 205–214.

Kohyama, K., and Nishinari, K. (1993). Dependence of the specific volume of konjac glucomannan on pH. In *Gums and Stabilizers for the Food Industry*, Phillips, G. O., Williams, P. A., and Wedlock, D. J. (Eds.), Vol. 5, pp. 459–462. IRL Press, Oxford.

Kohyama, K., Iida, H., and Nishinari, K. (1993). A mixed system composed of different molecular weights konjac glucomannan and kappa carrageenan: large deformation and dynamic viscoelastic study. *Food Hydrocoll.* 7:213–226.

Kokini, J. L., Lai, L-S., and Chedid, L. L. (1992). Effect of starch structure on starch rheological properties. *Food Technol.* 46:124–126, 128, 130, 132, 134–136, 138–139.

Konno, A., Azechi, Y., and Kimura, H. (1979). Properties of curdlan gel. *Agric. Biol. Chem.* 43:101–104.

Kovacs, P. (1973). Useful incompatibility of xanthan gum with galactomannans. *Food Technol.* 27:26–30.

Kowkabany, G. N., Binkley, W. W., and Wolfrom, M. L. (1953). Amino acids in cane juice and cane final molasses. *J. Agric. Food Chem.* 1:84–87.

Krag-Anderson, S., and Solderberg, J. R. (1992). A new method to determine the gelling temperature of food gels. In *Gums and Stabilizers for the Food Industry*, Phillips, G. O., Williams, P. A., and Wedlock, D. J. (Eds.), Vol. 6, pp. 117–120. IRL Press, Oxford.

Kravtchenko, T. P., Penci, M., Voragen, A. G. J., and Pilnik, W. (1993). Enzymic and chemical degradation of some industrial pectins. *Carbohydr. Polym.* 20:195–205.

Kravtchenko, T. P., Voragen, A. G. J., and Pilnik, W. (1992). Studies on the intermolecular distribution of industrial pectins by means of preparative ion-exchange chromatography. *Carbohydr. Polym.* 19:115–124.

Krieger, J. (1990). Bacterial cellulose near commercialization. *Chem. Eng. News* May 21:35–37.

Krumel, K. L., and Sarkar, N. (1975). Flow properties of gums useful to the food industry. *Food Technol.* 29:36–43.

Kubal, J. V., and Gralén, N. (1948). Physicochemical properties of karaya gum and locust bean mucilage. *J. Colloid Sci.* 3:457–471.

Kulicke, W. M., and Haas, R. (1984). Flow behavior of dilute polyacrylamide solutions through porous media. 1. Influence of chain length, concentration, and thermodynamic quality of solvent. *Ind. Eng. Chem. Fund.* 23:308–315.

Kulicke, W.-M., and Nottelmann, H. (1989). Structure and swelling of some synthetic, semisynthetic, and biopolymer hydrogels. In *Polymers in Aqueous Media*, Glass, J. E. (Ed.), Advances in Chemistry 223, pp. 15–44. Am. Chem. Soc., Washington, DC.

LaBell, F. (1991). New ingredients for low-fat, sugar-free, and microwave foods. *Food Processing* 52:52, 54

Lapasin, R., and Pricl, S. (1995). *Rheology of Industrial Polysaccharides: Theory and Applications.* Blackie, Glasgow/London.

Launay, B., Doublier, J. L., and Cuvelier, G. (1986). Flow properties of aqueous solutions and dispersions of polysaccharides. In *Functional Properties of Food Macromolecules*, Mitchell, J. R., and Ledward, D. A. (Eds.), pp. 1–78. Elsevier, Amsterdam/New York.

Launer, H. F., and Tomimatsu, Y. (1959). Reaction of sodium chlorite with various polysaccharides. *Anal. Chem.* 31:1569–1574.

Laurent, T. C., Preston, B. N., and Carlsson, B. (1974). Interaction between polysaccharides and other macromolecules. Conformational transitions of polynucleotides in polymeric media. *Eur. J. Biochem.* 43:231–235.

Leathers, T. D. (1993). Enzymatic saccharification of corn fiber. 206th National Meeting, Am. Chem. Soc., Div. Agric. Food Chem., Chicago, IL, Abstract 44.

Lechert, H., Maiwald, W., Köthe, R., and Basler, W-D. (1980). NMR-study of water in some starches and vegetables. *J. Food Processing and Preservation* 3:275–299.

Lee, J. C., and Lee, L. L. Y. (1979). Interaction of calf brain tubulin with poly(ethylene glycols). *Biochemistry* 18:5518–5526.

Lee, J. T., and Woo, K. L. (1988). Effect of preventing the Maillard reaction between casein and glucose with corn syrup and sucrose. *Korean J. Food Sci. and Technol.* 20:526–535 (*Food Sci. Technol. Abstr.* 5A13, 1989).

Lee, M. L., and Markides, K. E. (1987). Supercritical fluid chromatography. *Nature* (*London*) 327:441, 442; Chromatography with supercritical fluids. *Science* 235:1342–1347.

Lelièvre, J. (1992). Thermal analysis of carbohydrates as illustrated by aqueous starch systems. In *Developments in Carbohydrate Chemistry*, Alexander, R. J., and Zobel, H. F. (Eds.), pp. 137–161. Amer. Assoc. Cereal Chem., St. Paul, MN.

Leloup, V. M., Colonna, P., Ring, S. G., Roberts, K., and Wells, B. (1992). Microstructure of amylose gels. *Carbohydr. Polym.* 18:189–197.

Leonard, R. H. (1956). Levulinic acid as a basic chemical raw material. *Ind. Eng. Chem.* 48:1331–1341.

Levine, H., and Slade, L. (1988). Principles of "cryostabilization" technology from structure/property relationships of carbohydrate/water systems—a review. *Cryo-Lett.* 9:21–63.

Levine, H., and Slade, L. (1992). Glass transitions in foods. In *Physical Chemistry of Foods*, Schwartzberg, H. G., and Hartel, R. W. (Eds.), pp. 83–221. Dekker, New York.

Liebman, S. A., and Levy, E. J. (1983). Pyrolysis gas chromatography, mass spectrometry, and Fourier transform IR spectroscopy. In *Polymer Characterization*, Craver, C. D. (Ed.), Advances in Chemistry 203, pp. 617–634. Am. Chem. Soc., Washington, DC.

Lillford, P. J. (1986). Texturization of proteins. In *Functional Properties of Food Macromolecules*, Mitchell, J. R., and Ledward, D. A. (Eds.), pp. 355–384. Elsevier, Amsterdam.

Lin, C. F. (1977). Interaction of sulfated polysaccharides with proteins. In *Food Colloids*, Graham, H. D. (Ed.), pp. 320–346. AVI Press, New York.

Lips, A., Campbell, I. J., and Pelan, E. G. (1991). Aggregation mechanisms in food colloids and the role of biopolymers. In *Food Polymers, Gels and Colloids*, Dickinson, E. (Ed.), pp. 1–21. Royal Chem. Soc., London.

Lips, A., Hart, P. M., and Clark, A. H. (1988). Studies of biopolymer gels by compressive de-swelling. In *Gums and Stabilizers in the Food Industry*, Phillips, G. O., Williams, P. A., and Wedlock, D. J. (Eds.), Vol. 4, pp. 39–49. IRL Press, Oxford.

Lipska, A. E., and Parker, W. J. (1966). Kinetics of the pyrolysis of cellulose in the temperature range 250–300°C. *J. Appl. Polym. Sci.* 10:1439–1453.

Liu, H., and Lelièvre, J. (1992). A differential scanning calorimetry study of melting transitions in aqueous suspensions containing blends of wheat and rice starch. *Carbohydr. Polym.* 17:145–149.

Lopes, L., Andrade, C. T., Milas, M., and Rinaudo, M. (1992). Role of conformation and acetylation of xanthan on xanthan-guar interaction. *Carbohydr. Polym.* 17:121–126.

Lopes da Silva, J. A., Gonçalves, M. P., and Rao, M. A. (1993). Viscoelastic behavior of mixtures of locust bean gum and pectin dispersions. *J. Food Eng.* 18:211–228.

Luijkx, G. C. A., Vinke, P., van Rantwijk, F., and van Bekkum, H. (1991). Some Conversion Reactions of 5-Hydroxymethylfurfural, Abstract 58, Carbohydr. Div., Am. Chem. Soc., Washington, DC.

Lund, D. (1984). Influence of time, temperature, moisture, ingredients, and processing conditions on starch gelatinization. *Crit. Rev. Food Sci. Nutr.* 20:249–273.

Lupescu, N., Solo-Kwan, J., Christiaen, D., Morvan, H., and Arad, S. M. (1992). Structural determination by means of gas chromatography–mass spectrometry of 3-O-(α-D-glucopyranosyluronic acid)-galactopyranose, an aldobiuronic acid derived from *Porphyridium sp.* polysaccharide. *Carbohydr. Polym.* 19:131–134.

Mackie, A. R., Mingins, J., Dann, R., and North, A. N. (1991). Preliminary studies of β-lactoglobulin adsorbed on polystyrene latex. In *Food Polymers, Gels and Colloids*, Dickinson, E. (Ed.), pp. 96–112. Royal Chem. Soc., London.

Magee, R. J., and Kosaric, N. (1985). Bioconversion of hemicellulose. *Adv. Biochem. Eng. Biotechnol.* 32:61–93.

Major, W. D. (1958). The degradation of cellulose in oxygen and nitrogen at high temperatures. *Tech. Assoc. Pulp Paper Inst.* 41:530–536.

Malkki, Y., Heinio, R.-L., and Autio, K. (1993). Influence of oat gum, guar gum and carboxymethyl cellulose on the perception of sweetness and flavour. *Food Hydrocoll.* 6:525–532.

Mangat, M. N. (1992). Proprietary pectin-based fat replacer. *Food Technol.* 46:15154.

Marchessault, R. H., and Deslandes, Y. (1979). The structure of (1,3)-β-D-glucans: curdlan and paramylon. *Carbohydr. Res.* 75:231–242.

Marrinan, H. J., and Mann, J. (1956). Infrared spectra of the crystalline modifications of cellulose. *J. Polym. Sci.* 21:301–311.

Marrs, W. M. (1988). The effect of gamma radiation on the structure of carrageenans. In *Gums and Stabilizers for the Food Industry*, Phillips, G. O., Williams, P. A., and Wedlock, D. A. (Eds.), Vol. 4, pp. 399–408. IRL Press, Oxford.

Marsh, C. A. (1966). Chemistry of D-glucuronic acid and its glycosides. In *Glucuronic Acid: Free and Combined*, Dutton, G. J. (Ed.), pp. 71–77. Academic Press, New York.

Matsuhashi, T. (1990). Agar. In *Food Gels*, Harris, P. (Ed.), pp. 1–51. Elsevier, Amsterdam/New York.

Mayer, J. M. (1993). Processability and characterization of starch-blend films produced using different starch sources and amylose:amylopectin ratios. 206th National Meeting, Am. Chem. Soc., Div. Food & Agric. Sci., Chicago, IL, Abstract 129.

McCormick, C. L., and Shen, T. S. (1982). Cellulose dissolution and derivatization in lithium chloride/*N,N*-dimethylacetamide solutions, Seymour, R. B., and Stahl, G. A. (Eds.), Pergamon, New York.

McCurdy, R. D., Goff, H. D., Stanley, D. W., and Stone, A. P. (1994). Rheological properties of dextran related to food applications. *Food Hydrocoll.* 8:609–623.

McIntyre, D. D., and Vogel, H. J. (1993). Structural studies of pullulan by nuclear magnetic resonance. *Starch/Staerke* 45:406–410.

McNamara, M. K., and Stone, B. A. (1983). Isolation, characterization and chemical synthesis of a galactosyl-hydroxyproline linkage compound from wheat endosperm arabinogalactan-peptide. *Lebensm. Wiss. Technol.* 14:182–187.

McReady, R. M., and Reeve, R. M. (1955). Test for pectin based on reaction of hydroxamic acids with ferric iron. *J. Agric. Food Chem.* 3:260–262.

Meer, W. (1980a). Agar. In *Handbook of Water-Soluble Gums and Resins*, Davidson, R. L. (Ed.), pp. 7-1–7-19. McGraw-Hill, New York.

Meer, W. (1980b). Gum arabic. In *Handbook of Water-Soluble Gums and Resins*, Davidson, R. L. (Ed.), pp. 8-1–8-24. McGraw-Hill, New York.

Meer, W. (1980c). Gum karaya. In *Handbook of Water-Soluble Gums and Resins*, Davidson, R. L. (Ed.), pp. 10-1–10-14. McGraw-Hill, New York.

Mehl, J. W., Oncley, J. L., and Simha, R. (1940). Viscosity and the shape of protein molecules. *Science* 92:132ff.

Mehltretter, C. L. (1964). Dialdehyde starch. *Methods in Carbohydrate Chemistry*, Vol. IV, p. 316. Academic Press, New York.

Merritt, C., Jr., and Angelini, P. (1971). Formation of flavour compounds from proteins and amino acids. Meeting, Am. Chem. Soc., Agric. Food Chem. Div., September 12–17, Washington, DC, Abstract 162.

Merritt, C., Jr., and Robertson, D. H. (1967). The analysis of proteins, peptides and amino acids by pyrolysis–gas chromatography and mass spectrometry. *J. Gas Chromatogr.* 5:96–98.

Metko, S. K., and McFeeters, R. F. (1993). Effect of the degree of methylation and pH on nonenzymatic degradation of pectin. 206th National Meeting, Am. Chem. Soc., Div. Agric. & Food Chem., Chicago, IL, Abstract 112.

Michel, F., Thibault, J.-F., and Doublier, J.-L. (1984). Viscometric and potentiometric study of high-methoxyl pectins in the presence of sucrose. *Carbohydr. Polym.* 4:283–297.

Miles, M. J., Morris, V. J., Orford, P. D., and Ring, S. G. (1985). The roles of amylose and amylopectin in the gelation and retrogradation of starch. *Carbohydr. Res.* 135:271–281.

Miles, M. J., Tanaka, K., and Keller, A. (1983). The behaviour of polyelectrolyte solutions in elongational flow; the determination of conformational relaxation times (with an Appendix of an anomalous adsorption effect). *Polymer* 24:1081–1088.

Mita, T. (1992). Structure of potato starch pastes in the ageing process by the measurement of their dynamic moduli. *Carbohydr. Polym.* 17:269–276.

Mitchell, J. R. (1979). Rheology of polysaccharide solutions and gels. In *Polysaccharides in Food*, Blanshard, J. M. V., and Mitchell, J. R. (Eds.), pp. 54, 55. Butterworth, London.

Mitchell, J. R., and Blanshard, J. M. V. (1979). On the nature of the relationship between the structure and rheology of food gels. In *Food Texture and Rheology*, Sherman, P. (Ed.), pp. 425–435. Academic Press, New York.

Mitchell, J. R., Reed, J., Hill, S. E., and Rogers, E. (1991). Systems to prevent loss of functionality on heat treatment of galactomannans. *Food Hydrocoll.* 5:141–143.

Miyawaki, M., and Konno, A. (1982). Citrus Food Containing a Cyclodextrin. U.S. Patent No. US 4 332 825.

Miyoshi, E., Takaya, T., and Nishinari, K. (1994). Gel-sol transition in gellan gum solutions. I. Rheological studies on the effects of salt; II. DSC studies on the effects of salt. *Food Hydrocoll.* 8:505–542.

Mohsenin, N. N. (1980). *Physical Properties of Plant and Animal Materials*, p. 173. Gordon and Breach, New York.

Moirano, A. L. (1977). Sulfated seaweed polysaccharides. In *Food Hydrocolloids*, Graham, H. D. (Ed.), pp. 347–381. AVI Press, New York.

Montgomery, E. M., and Senti, F. R. (1958). Separation of amylose from amylopectin of starch by an extraction–sedimentation procedure. *J. Polym. Sci.* 28:1–9.

Moon, M. H., and Giddings, J. C. (1993). Rapid separation and measurement of particle size distribution of starch granules by sedimentation/steric field-flow fractionation. *J. Food Sci.* 58:1166–1171.

Morita, H. (1956). Characterization of starch and related polysaccharides by differential thermal analysis. *Anal. Chem.* 28:64–67.

Morita, H. (1957). Differential thermal analysis of some polyglucosans. *Anal. Chem.* 29: 1095–1097.

Morita, N., Hayashi, K., Takagi, M., and Miyano, K. (1983). Gas–liquid chromatography and mass spectrometry of quinoxalines derived from various homoglucans by alkaline *o*-phenylenediamine method. *Agric. Biol. Chem.* 47:757–763.

Moritaka, H., Nishinari, K., Horiuchi, H., and Watase, M. (1980). Rheological properties of aqueous agarose-gelatin gels. *J. Text. Studies* 11:257–270.

Morgan, K. R., Furneaux, R. H., and Stanley, R. A. (1992). Observation by solid-state -3C CP MAS NMR spectroscopy of the transformations of wheat starch associated with the making and staling of bread. *Carbohydr. Res.* 235:15–22.

Morris, E. R. (1976). Molecular origin of xanthan solution properties. 172nd National Meeting, Am. Chem. Soc., Carbohydr. Div., San Francisco, CA, Abstract 38.

Morris, E. R. (1987). Organoleptic properties of food polysaccharides in thickened systems. In *Industrial Polysaccharides*, Yalpani, M. (Ed.), pp. 225–238. North-Holland, Amsterdam.

Morris, E. R. (1990). Mixed polymer gels. In *Food Gels*, Harris, P. (Ed.), pp. 291–359. Elsevier, Amsterdam/New York.

Morris, E. R. (1994). Chiroptical methods. In *Physical Techniques for the Study of Food Biopolymers*, Ross-Murphy, S. B. (Ed.), pp. 15–64. Blackie, Glasgow/London.

Morris, E. R., and Norton, I. T. (1983). Polysaccharide aggregation in solution and gels. In *Aggregation Process in Solution*, Wyn-Jones, E., and Gormally, J. (Eds.). Elsevier, Amsterdam/New York, 1983.

Morris, E. R., Cutler, A. N., Ross-Murphy, S. B., Rees, D. A., and Price, J. (1981). Concentration and shear rate dependence of viscosity in random coil polysaccharide solutions. *Carbohydr. Polym.* 1:5–21.

Morris, E. R., Gidley, M. J., Murray, E. J., Powell, D. A., and Rees, D. A. (1980). Characterization of pectin gelation under conditions of low water activity, by circular dichroism, competitive inhibition and mechanical properties. *Int. J. Biol. Macromol.* 2:327–330.

Morris, E. R., Rees, D. A., and Robinson, G. (1980). Cation-specific aggregation of carrageenan helices: domain model of polymer gel structure. *J. Mol. Biol.* 138:349–362.

Morris, E. R., Rees, D. A., Young, G., Walkinshaw, M. D., and Drake, A. (1977). Order–disorder transitions for a bacterial polysaccharide. *J. Mol. Biol.* 110:1–16.

Morris, V. J. (1985). Food gels–roles played by polysaccharides. *Chem. Ind.* (*London*) March, pp. 159–164.

Morris, V. J. (1992). Designing polysaccharides for synergistic interactions. In *Gums and Stabilizers for the Food Industry*, Phillips, G. O., Williams, P. A., and Wedlock, D. J. (Eds.), Vol. 6, pp. 161–171. IRL Press, Oxford.

Morris, V. J., and Chilvers, G. R. (1984). Cold setting alginate-pectin mixed gels. *J. Sci. Food Agric.* 35:1370–1376.

Morrison, I. A. (1975). Determination of the degree of polymerization of oligo- and polysaccharides by gas-liquid chromatography. *J. Chromatogr.* 108:361–364.

Morrison, W. R., and Scott, D. C. (1986). Measurement of the dimensions of wheat starch granule population using a Coulter counter with 100-channel analyzer. *J. Cereal Sci.* 4:13–21.

Morrison, W. R., Law, R. V., and Snape, C. E. (1993). Evidence for inclusion complexes of lipids with V-amylose in maize, rice and oat starches. *J. Cereal Sci.* 18:107–109.

Muyldermans, G. (1993). The use of blends as fat mimetics: gelatin/hydrocolloid combinations. In *Handbook of Fat Replacers*, Roller, S., and Jones, S. A. (Eds.), pp. 251–263. CRC Press, Boca Raton, FL.

Munk, P. (1989). *Introduction to Macromolecular Science*. Wiley, New York.

Myers, R. R. (1960). The physical stability of dispersions. In *Physical Functions of Hydrocolloids*. Advances in Chemistry 25, pp. 92–103. Am. Chem. Soc., Washington, DC.

Nadison, J. (1990). The interaction of carrageenan and starch in cream desserts. *Dairy Ind. Int.* 55:12–13.

Nakanishi, I., Kimura, K., Kusui, S., and Yamazaki, E. (1974). Complex formation of gel-forming bacterial (1,3)-β-D-glucans (curdlan-type polysaccharides) with dyes in aqueous solution. *Carbohydr. Res.* 32:47–52.

Nakao, Y., Konno, A., Taguchi, T., Tawada, T., Kasai, H., Toda, J., and Terasaki, M. (1991). Curdlan: properties and application to foods. *J. Food Sci.* 56:769–776.

Nelson, K. L., and Fennema, O. R. (1991). Methylcellulose films to prevent lipid migration in confectionery products. *J. Food Sci.* 56:504–509.

Neukom, H. (1976). Chemistry and properties of the non-starchy polysaccharides (nsp) of wheat flour. *Lebensm. Wiss. Technol.* 9:143–148.

Neukom, H., and Markwalder, H. U. (1978). Oxidative gelation of wheat flour pentosans: a new way of cross-linking polymers. *Cereal Foods World* 23:374–376.

Nichols, J. B., and Bailey, E. D. (1949). Determinations with the ultracentrifuge. In *Physical Methods of Organic Chemistry*, Weissberger, A. (Ed.), p. 629. Interscience, New York.

Nicoll, W. D., and Conaway, R. F. (1943). Alkali and other metal derivatives. In *Cellulose and Cellulose Derivatives*, Ott, E. (Ed.), pp. 709–757. Interscience, New York.

Nishinari, K., and Watase, M. (1992). Effects of sugars and polyols on the gel–sol transition of kappa-carrageenan gels. *Thermochim. Acta* 206:149–162.

Nishinari, K., Watase, M., Kohyama, K., and Moritaka, H. (1992). DSC study on the gel–sol transition of mixed gels of agarose-gelatin, kappa-carrareenan and funoran-gelatin. In *Gums and Stabilizers for the Food Industry*, Phillips, G. O., Williams, P. A., and Wedlock, D. J. (Eds.), Vol. 6, pp. 191–200. IRL Press, Oxford.

Nishinari, K., Watase, M., Kohyama, K., Nishinari, N., Oakenfull, D., Koide, S., Ogino, K., Williams, P. A., and Phillips, G. O. (1992). The effect of sucrose on the thermo-reversible gel–sol transition in agarose and gelatin. *Polym. J.* 24:871–877.

Noel, T. R., Ring, S. G., and Whittam, M. A. (1993). Physical properties of starch products: structure and function. In *Food Colloids and Polymers: Stability and Mechanical Properties*, Dickinson, E., and Walstra, P. (Eds.), pp. 126–137. Royal Chem. Soc., London.

Norman, A. G. (1943). Carbohydrates normally associated with cellulose in nature. In *Cellulose and Cellulose Derivatives*, Ott, E. (Ed.), p. 433. Interscience, New York.

Oakenfull, D. G. (1991). The chemistry of high-methoxyl pectins. In *The Chemistry and Technology of Pectin*, Walter, R. H. (Ed.), pp. 87–105. Academic Press, New York.

Oakenfull, D., and Sidhu, G. S. (1992). Low-cholesterol egg and dairy products. *Outlook Agric.* 21:203–208.

Oates, C. G., Lucas, P. W., and Lee, W. P. (1993). How brittle are gels? *Carbohydr. Polym.* 20:189–194.

Odell, J. A., Keller, A., and Müller, A. J. (1989). Extensional flow behavior of macromolecules in solution. In *Polymers in Aqueous Media*, Glass, J. E. (Ed.), Advances in Chemistry 223, pp. 193–244. Am. Chem. Soc., Washington, DC.

Ogawa, K., Tsurugi, J., and Watanabe, T. (1973a). The dependence of the conformation of a β-1,3-D-glucan on chain length in alkaline solution. *Carbohydr. Res.* 29:397–403.

Ogawa, K., Tsurugi, J., and Watanabe, T. (1973b). Effect of salt on the conformation of gel-forming β-1,3-D-glucan in alkaline solution. *Chemistry Lett.* 95–98.

Ogawa, K., Watanabe, T., Tsurugi, J., and Ono, S. (1972). Conformational behavior of a gel-forming (1,3)β-D-glucan in alkaline solution. *Carbohydr. Res.* 23:399–405.

Ogston, A. G., and Stainier, J. E. (1951). The dimension of the particles of hyaluronic acid complex in synovial fluid. *Biochem. J.* 49:585–590.

Okechukwu, P. E., and Rao, M. A. (1996a). Kinetics of cowpea starch gelatinization based on granule swelling. *Starch/Starke* 48:43–47.

Okechukwu, P. E., and Rao, M. A. (1996b). Food Hydrocoll., to appear.

Okiyama, A., Motoki, M., and Yamanaka, S. (1993). Bacterial cellulose III. Development of a new form of cellulose. *Food Hydrocoll.* 6:493–501.

Olkku, J. (1978). Gelatinization of starch and wheat flour starch—a review. *Food Chem.* 3:293–317.

Oppermann, W. (1992). Swelling behavior and elastic properties of ionic hydrogels. In *Polyelectrolyte Gels*, Harland, R. S., and Prud'homme, R. K. (Eds.), ACS Symposium Series 480, pp. 159–170. Am. Chem. Soc., Washington, DC.

Osman, E. M. (1967). Starch in the food industry. In *Starch: Chemistry and Technology*, Whistler, R. L., and Paschall, E. F. (Eds.), Vol. 2, p. 169. Academic Press, New York.

Oster, G. (1960). Light scattering. In *Physical Methods of Organic Chemistry*, Weissberger, A. (Ed.), pp. 2107–2145. Interscience, New York.

Ostgaard, K. (1992). Enzymic microassay for the determination and characterization of alginates. *Carbohydr. Polym.* 19:51–59.

Ott, E. (1943). *Cellulose and Cellulose Derivatives*. Interscience, New York.

Oxford, A. E. (1945). The chemistry of antibiotic substances other than penicillin. In *Annual Reviews in Biochemistry*, Luck, J. M. (Ed.), p. 765. Annual Reviews, Palo Alto.

Pals, D. T. F., and Hermans, J. J. (1952). Sodium salts of pectin and of carboxy methyl cellulose in aqueous sodium chloride. *Rec. Trav. Chim. Pays-Bas* 71:433–467.

Panchev, I. N., Kirtchev, N. A., and Kratchanov, C. G. (1994). On the production of low esterified pectins by acid maceration of pectic raw materials with ultrasound treatment. *Food Hydrocoll.* 8:9–17.

Papageorgiou, M., Kasapis, S., and Richardson, R. K. (1994). Steric exclusion phenomena in gellan/gelatin systems I. Physical properties of single and binary gels. *Food Hydrocoll.* 8:97–112.

Parrish, M. A. (1987). Cyclodextrins—A review. *Specialty Chemicals* 7:366, 370, 372, 374, 378–380.

Pastor, M. V., Costell, E., Izquierdo, L., and Duran, L. (1994). Effects of concentration, pH and salt content on flow characteristics of xanthan gum solutions. *Food Hydrocoll.* 8:265–275.

Patel, P. D., and Hawes, G. B. (1988). Estimation of food-grade galactomannans by enzyme-linked lectin assay. *Food Hydrocoll.* 2:107–118.

Pechanek, U., Blaicher, G., Pfannhauser, W., and Woidich, H. (1982). Electrophoretic method for qualitative and quantitative analysis of gelling and thickening agents. *J. Assoc. Offic. Anal. Chem.* 65:745–752.

Pecora, R. (Ed.) (1986). *Dynamic Light Scattering*. Plenum, New York.

Penichter, K. A., and McGinley, E. J. (1991). Cellulose gel for fat-free food applications. *Food Technol.* 45:105.

Pettitt, D. J. (1982). Xanthan gum. In *Food Hydrocolloids*, Glicksman, M. (Ed.), Vol. 1, pp. 127–149. CRC Press, Boca Raton, FL.

Pfannemüller, B., and Bauer-Carnap, A. (1977). Electron microscope studies on fibrils formed from retrograded synthetic amyloses. *Colloid Polym. Sci.* 255:844–848.

Pfizer, Inc. (1991). Litesse. *Food Technol.* 45:102ff.

Pilgrim, G. W., Walter, R. H., and Oakenfull, D. G. (1991). Jams, jellies and preserves. In *The Chemistry and Technology of Pectin*, Walter, R. H. (Ed.), pp. 23–50. Academic Press, San Diego, CA.

Pippen, E. L., McCready, R. M., and Owens, H. S. (1950). Gelation properties of partially acetylated pectins. *J. Am. Chem. Soc.* 72:813–816.

Platt, W., and Fleming, R. S. (1923). The action of shortening in the light of the newer theories of surface phenomena. *Ind. Eng. Chem.* 15:390–394.

Poland, D., and Scheraga, H. A. (1970). *Theory of Helix-Coil Transitions in Biopolymers*. Academic Press, New York.

Ponder, G. R., and Richards, G. N. (1993). Pyrolysis of inulin, glucose and fructose. *Carbohydr. Res.* 244:341–359.

Poovaiah, B. W., and Moulton, G. A. (1982). Vacuum Pressure Infiltration Process for Fresh Produce. U.S. Patent 4 331 691.

Powell, R. E., and Eyrin, H. (1942). Frictional and thermodynamic properties of large molecules. *Advances in Colloid Science*, Vol. 1, p. 186. Interscience, New York.

Praznik, W., Beck, R. H. F., and Nitsch, E. (1984). Determination of fructan oligomers of degree of polymerization 2–30 by high-performance liquid chromatography. *J. Chromatogr.* 303: 417–421.

Provder, T., Holsworth, R. M., Grentzer, T. H., and Kline, S. A. (1983). Use of the single dynamic temperature scan method in differential scanning calorimetry for quantitative reaction kinetics. In *Polymer Characterization*, Craver, C. D. (Ed.), *Advances in Chemistry* 203, pp. 233–253. Am. Chem. Soc., Washington, DC.

Prud'homme, R. K., Constien, V., and Knoll, S. (1989). The effects of shear history on the rheology of hydroxypropyl guar gels. In *Polymers in Aqueous Media*, Glass, J. E. (Ed.), *Advances in Chemistry* 223, pp. 90–112. Am. Chem. Soc., Washington, DC.

Pszczola, D. E. (1991). Carbohydrate-based ingredient performs like a fat for use in a variety of food applications. *Food Technol.* 45:262, 263, 276.

Pucher, G. W., Leavenworth, C. S., and Vickery, H. B. (1948). Determination of starch in plant tissues. *Anal. Chem.* 20:850–853.

Purves, C. B. (1943). Chemical nature of cellulose and its derivatives. In *Cellulose and Cellulose Derivatives*, Ott, E. (Ed.), pp. 54, 110. Interscience, New York.

Qu, D., and Wang, S. S. (1991a). Kinetics of formation of gelatinized and melted starches at extrusion-cooking conditions. IFT Annual Meeting, Chicago, IL, Abstract 43.

Qu, D., and Wang, S. S. (1991b). Properties of gelatinized and melted starch and their separation from extrusion-cooked product. IFT Annual Meeting, Chicago, IL, Abstract 42.

Quenin, I., and Chanzy, H. (1987). Polymorphic and morphological aspects of recrystallized cellulose as a function of molecular weight. In *The Structures of Cellulose*, Attala, R. H. (Ed.), ACS Symposium Series 340, pp. 189–198. Am. Chem. Soc., Washington, DC.

Radosta, S., Schierbaum, F., and Yuriev, W. P. (1989). Polymer–water interaction of maltodextrins. II. NMR study of bound water in liquid maltodextrin–water systems. *Starch/Staerke* 41:428–430.

Raemy, A., and Schweizer, T. F. (1983). Thermal behavior of carbohydrates studied by heat flow calorimetry. *J. Thermal Anal.* 28:95–107.

Randall, R. C., Phillips, G. O., and Williams, P. A. (1988). The role of the proteinaceous component on the emulsifying properties of gum arabic. *Food Hydrocoll.* 2:131–140.

Rao, M. A. (1992). Measurement of viscoelastic properties of fluid and semisolid foods. In *Viscoelastic Properties of Food*, Rao, M. A., and Steffe, J. F. (Eds.), pp. 207–231. Elsevier, Amsterdam/New York.

Rao, M. A., and Kenny, J. F. (1975). Flow properties of selected food gums. *Can. Inst. Food Sci. Technol. J.* 8:142–148.

Rao, M. A., Cooley, H. J., Walter, R. H., and Downing, D. L. (1989). Evaluation of texture of pectin jellies with the Voland–Stevens texture analyzer. *J. Text. Studies* 20:87–95.

Rao, M. A., Walter, R. H., and Cooley, H. J. (1981). Effect of heat treatment on the flow properties of aqueous guar gum and sodium carboxymethylcellulose (CMC) solutions. *J. Food Sci.* 46:896–902.

Rao, V. N. M. (1992). Viscoelastic properties of solid foods. In *Viscoelastic Properties of Foods*, Rao, M. A., and Steffe, J. F. (Eds.), pp. 3–47. Elsevier, Amsterdam/New York.

Rees, D. A. (1969). Structure, conformation, and mechanism in the formation of polysaccharide gels and networks. *Adv. Carbohydr. Chem. Biochem.* 24:267–332.

Rees, D. A. (1972a). Shapely polysaccharides. *Biochem. J.* 126:257–273.

Rees, D. A. (1972b). Mechanism of gelation in polysaccharide systems. In *Gelation and Gelling Agents*, Symposium Proceedings, 13, pp. 7–12. British Food Manufacturing Industry Research Association, London.

Rees, D. A. (1972c). Polysaccharide gels. *Chem. Ind. (London)* 630–636.

Rees, D. A., and Wight, A. W. (1971). Polysaccharide conformation. Part VII. Model building computations for α-1,4 galacturonan and the kinking function of L-rhamnose residues in pectic substances. *J. Chem. Soc. B* 1366–1372.

Rees, D. A., Morris, E. R., Thom, D., and Madden, J. K. (1982). Shapes and interactions of carbohydrate chains. In *The Polysaccharides*, Aspinall, G. O. (Ed.), pp. 195–290. Academic Press, New York.

Reeves, R. E., and Blouin, F. A. (1957). The shape of pyranoside rings. II. The effect of sodium hydroxide upon the optical rotation of glycosides *J. Am. Chem. Soc.*, 79:2261–2264.

Reichman, D., and Garti, N. (1991). Galactomannans as emulsifiers. In *Food Polymers, Gels and Colloids*, Dickinson, E. (Ed.), pp. 549–556. Royal Chem. Soc., London.

Reineccius, G. A. (1991). Carbohydrates for flavor encapsulation. *Food Technol.* 45:144–146.

Reinhart, G. D. (1980). Influence of polyethylene glycols on the kinetics of rat liver phosphofructokinase. *J. Biol. Chem.* 255:10576–10578.

Reisenhofer, E., Cesaro, A., Delben, F., Manzini, G., and Paoletti, S. (1984). Copper(II) binding by natural ionic polysaccharides. Part II. Polarographic data. *Bioelectrochem. Bioenerg.* 12: 455–465 (*Chem. Abstr.* 102:149687, 1985).

Reiter, F. (1986). Films, cellophane. Packaging Reference Issue, p. 52f.

Rexova-Benkova, L., and Markovic, O. (1976). Pectic enzymes. *Adv. Carbohydr. Chem.* 33:323–385.

Richards, G. N. (1963). Alkaline degradation. *Methods in Carbohydrate Chemistry*, Vol. III, pp. 154–164. Academic Press, New York.

Richards, G. N., and Streamer, M. (1972). Studies on dextranases. Part I. Isolation of extracellular, bacterial dextranases. *Carbohydr. Res.* 25:323–332.

Richmond, M. D., and Yeung, E. S. (1993). Development of laser-excited indirect fluorescence detection for high-molecular-weight polysaccharide in capillary electrophoresis. *Anal. Biochem.* 210:245–248.

Rickayzen, G. (1989). Theoretical studies of the solid-fluid interface. In *Food Colloids*, Bee, R. D., Richmond, P., and Mingins, J. (Eds.), Special Publication series 75, p. 153. Royal Chem. Soc., London.

Rinaudo, M. (1988). Gelation of ionic polysaccharides. In *Gums and Stabalizers for the Food Industry*, Phillips, G. O., Williams, P. A., and Wedlock, D. J. (Eds.), Vol. 4, pp. 119–126. IRL Press, Oxford.

Rinaudo, M. (1992). The relation between the chemical structure of polysaccharides and their physical states. In *Gums and Stabilizers for the Food Industry*, Phillips, G. O., Williams, P. A., and Wedlock, D. J. (Eds.), Vol. 6, p. 52. IRL Press, Oxford.

Ring, G. J. F. (1982). A study of the polymerization kinetics of bacterial cellulose through gel-permeation chromatography. In *Cellulose and Other Natural Polymer Systems*, Brown, R. M., Jr. (Ed.), pp. 299–325. Plenum, New York.

Ring, S. G., and Whittam, M. A. (1991). Linear dextrins. In *Biotechnology of Amylodextrin Oligosaccharides*, Friedman, R. B. (Ed.), ACS Symposium Series 458, p. 281. Am. Chem. Soc., Washington, DC.

Ring, S. G., Miles, M. J., Morris, V. J., Turner, R., and Colonna, P. (1987). Spherulitic crystallization of short chain amylose. *Int. J. Biol. Macromol.* 9:158–160.

Risch, S. J., and Reineccius, G. A. (Eds.) (1995). *Encapsulation and Controlled Release of Food Ingredients*, ACS Symposium Series 590. Am. Chem. Soc., Washington, DC.

Robb, I. D. (1986). Interaction of polymers. In *Gums and Stabilizers for the Food Industry*, Phillips, G. O., Wedlock, D. J., and Williams, P. A. (Eds.), Vol. 3, pp. 399–407. Elsevier, Amsterdam/New York.

Robinson, G., Manning, C. E., and Morris, E. R. (1991). Conformation and physical properties of the bacterial polysaccharides gellan, whelan, and rhamsan. In *Food Polymers, Gels and Colloids*, p. 31. Royal Chem. Soc., London.

Robinson, G., Manning, C. E., Morris, E. R., and Dea, I. C. M. (1988). Sidechain–mainchain interactions in bacterial polysaccharides. In *Gums and Stabilizers for the Food Industry*, Phillips, G. O., Williams, P. A., and Wedlock, D. J. (Eds.), Vol. 4, pp. 173–181. IRL Press, Oxford.

Robinson, G., Ross-Murphy, S. B., and Morris, E. R. (1982). Viscosity–molecular weight relationships, intrinsic chain flexibility, and dynamic solution properties of guar galactomannan. *Carbohydr. Res.* 107:17–32.

Rocks, J. K. (1971). Xanthan gum. *Food Technol.* 25:476–477.

Roesen, J. (1992). Rheological characterization of carrageenan gels. In *Gums and Stabilizers for the Food Industry*, Phillips, G. O., Williams, P. A., and Wedlock, D. J. (Eds.), Vol. 6, p. 121. IRL Press, Oxford.

Roller, S. (1996). Starch-derived fat mimetics: maltodextrins. In *Handbook of Fat Replacers*, Roller, S., and Jones, S. A. (Eds.), pp. 99–118. CRC Press, Boca Raton, FL.

Roller, S., and Jones, S. A. (Eds.) (1996). *Handbook of Fat Replacers*. CRC Press, Boca Raton, FL.

Rollings, J. E., Bose, A., Caruthers, J. M., Tsao, G. T., and Okos, M. R. (1983). Aqueous size exclusion chromatography. In *Polymer Characterization*, Advances in Chemistry 203, pp. 345–360. Am. Chem. Soc., Washington, DC.

Rombouts, F. M., and Thibault, J. F. (1986). Feruloylated pectic substances from sugar-beet pulp. *Carbohydr. Res.* 154:177–188.

Ross, S., and Morrison, I. D. (1988). *Colloidal Systems and Interfaces*. Wiley, New York.

Ross-Murphy, S. B. (1984). Rheological methods. In *Biophysical Methods in Food Research*, Chan, H. W.-S. (Ed.), p. 167. Soc. Chem. Ind., London.

Ross-Murphy, S. B. (1991). Concentration dependence of gelation time. In *Food Polymers, Gels and Colloids*, Dickinson, E. (Ed.), p. 357. Royal Chem. Soc., London.

Ross-Murphy, S. B. (1994). *Physical Techniques for the Study of Food Biopolymers*. Blackie, Glasgow/London.

Russell, T. P., Deline, V. R., Dozier, W. D., Felcher, G. P., Agrawal, G., Wool, R. P., and Mays, J. W. (1993). Direct observation of reptation at polymer interfaces. *Nature (London)* 365:235–237.

Rutenberg, M. W. (1980) Starch and its modifications. In *Handbook of Water-Soluble Gums and Resins*. Davidson, R. L. (Ed.), pp. 22-1–22-83. McGraw-Hill, New York.

Sajjaanantakul, T., Van Buren, J. P., and Downing, D. L. (1993). Effect of cations on heat degradation of chelator-soluble carrot pectin. *Carbohydr. Polym.* 20:207–214.

Sajjan, S. U., and Rao, M. R. R. (1989). Functional properties of native and carboxymethyl guar gum. *J. Sci. Food Agric.* 48:377–380.

Salamon, A. G. (1900). The manufacture of caramel. *J. Soc. Chem. Ind.* 19:301–307.

Samsel, E. P., and Aldrich, J. C. (1957). Application of anthrone test to determination of cellulose derivatives in non-aqueous media. *Anal. Chem.* 29:574–576.

Sander, P. A. (1992). Fat-simulating hydrolyzed rice solids. *Food Technol.* 46:152.

Sanderson, G. R. (1990). Gellan gum. In *Food Gels*, Harris, P. (Ed.), pp. 201–231. Elsevier, Amsterdam/New York.

Sanderson, G. R., Bell, V. L., Clark, R. C., and Ortega, D. (1988). The texture of gellan gum. In *Gums and Stabilizers for the Food Industry*, Phillips, G. O., Williams, P. A., and Wedlock, D. J. (Eds.), Vol. 4, pp. 219–229. IRL Press, Oxford.

Sandford, P. A. (1979). Exocellular microbial polysaccharides. *Adv. Carbohydr. Chem. Biochem.* 36:265–313.

Sanofi (1988). *Hydrocolloids*. Sanofi Bioindustries, Paris, France.

Sarkar, N. (1979). Thermal gelation properties of methyl and hydroxypropyl methylcellulose. *J. Appl. Polym. Sci.* 24:1073–1087.

Sawyer, L. H., and George, W. (1982). Comparisons between synthetic and natural microfiber systems. In *Cellulose and Other Natural Polymer Systems*, Brown, R. M., Jr. (Ed.), pp. 429–455. Plenum, New York.

Schaller, D. (1977). Analysis of dietary fiber. *Food Product Dev.* 77:70–72.

Schenz, T., and Fugitt, M. (1992). The rheology of thin liquids. In *Gums and Stabilizers for the Food Industry*, Phillips, G. O., Williams, P. A., and Wedlock, D. J. (Eds.), Vol. 6, pp. 113–116. IRL Press, Oxford.

Schmitz, K. S. (1990). *An Introduction to Dynamic Light Scattering by Macromolecules*. Academic Press, New York.

Schneeman, B. O. (1986). Dietary fiber: physical and chemical properties, methods of analysis, and physiological effects. *Food Technol.* 40:104–110.

Schoch, T. J. (1942). Fractionation of starch by selective precipitation with butanol. *J. Am. Chem. Soc.* 64:2957–2961.

Schoch, T. J. (1964). Determination of alkali number. *Methods in Carbohydrate Chemistry*, Vol. IV, p. 61. Academic Press, New York.

Schols, H. A., Reitsma, J. C. E., Voragen, A. G. J., and Pilnik, W. (1989). High-performance ion exchange chromatography of pectins. *Food Hydrocoll.* 3:115–121.

Scholte, T. G. (1975). Sedimentation techniques. In *Polymer Molecular Weights*, Part II, Slade, P. E., Jr. (Ed.), pp. 501–589. Dekker, New York.

Schulz, D., Rau, U., and Wagner, F. (1992). Characteristics of films prepared from native and modified branched β-1,3-D-glucans. *Carbohydr. Polym.* 18:295–299.

Schwartzberg, H. G., and Hartel, R. W. (1992). *Physical Chemistry of Foods*. IFT Basic Symposium Series, p. iii. Dekker, New York.

Scott, J. E. (1965). Fractionation by precipitation with quaternary ammonium salts. *Methods in Carbohydrate Chemistry*, Vol. V, pp. 38–44. Academic Press, New York.

Segre, A. L., and Capitani, D. (1993). NMR in solid polymers. *Trends Polym. Sci.* 1:280–284.

Severs, E. T. (1962). *Rheology of Polymers*. Reinhold, New York.

Seymour, R. B., and Carraher, C. E., Jr. (1981). *Polymer Chemistry*. Dekker, New York.

Seymour, R. B., Johnson, E. L., and Stahl, G. A. (1982). Preparation and properties of cellulose blends in dimethylsulfoxide solution. In *Macromolecular Solutions*, Seymour, R. B., and Stahl, G. A. (Eds.). Pergamon, Elmsford, NY.

Shafizadeh, F. (1968). Pyrolysis and combustion of cellulose materials. *Adv. Carbohydr. Chem.* 23:419, 474.

Shah, C. B., and Barnett, S. M. (1992). Hyaluronic acid gels. In *Polyelectrolyte Gels*, Harland, R. S., and Prud'homme, R. K. (Eds.), ACS Symposium Series 480, pp. 116–130. Am. Chem. Soc., Washington, DC.

Sharman, W. R., Richards, E. L., and Malcolm, G. N. (1978). Hydrodynamic properties of aqueous solutions of galactomannans. *Biopolymers* 17:2817–2833.

Shasha, B., and Whistler, R. L. (1964). End-group analysis by periodate oxidation. *Methods in Carbohydrate Chemistry*, Vol. IV, pp. 86–88. Academic Press, New York.

Shaw, D. J. (1992). *Introduction to Colloid and Surface Chemistry*, 4th ed. Butterworth-Heinemann, Oxford.

Shen, T. C., and Cabasso, I. (1982). Ethyl cellulose anisotropic membranes. In *Macromolecular Solutions*, Seymour, R. B., and Stahl, G. A. (Eds.), pp. 108–119. Pergamon, Elmsford, NY.

Shim, J. L. (1985). Gellan Gum/Gelatin Blends U.S. Patent 4 517 216.

Shimada, K., Kawano, K., Ishii, J., and Nakamura, T. (1992). Structure of inclusion complexes of cyclodextrins with triglycerides at vegetable/oil/water interface. *J. Food Sci.* 57:655–656.

Shomer, I., Frenkel, H., and Polinger, C. (1991). The existence of a diffuse electric double layer at cellulose fibril surfaces and its role in the swelling mechanism of parenchyma plant cell walls. *Carbohydr. Polym.* 16:199–210.

Sidebotham, R. L. (1974). Dextrans. *Adv. Carbohydr. Chem. Biochem.* 30:371–444.

Silberberg, A. (1989). Gelled aqueous systems. In *Polymers in Aqueous Media*, Glass, J. E. (Ed.), Advances in Chemistry 223, p. 5. Am. Chem. Soc., Washington, DC.

Silberberg, A. (1992). Gel structural heterogeneity, gel permeability, and mechanical response. In *Polyelectrolyte Gels*, Harland, R. S., and Prud'homme, R. K. (Eds.), ACS Symposium Series 480, pp. 146–158. Am. Chem. Soc., Washington, DC.

Silge, M. R. (1992). Dried-fruit-based fat replacement systems. *Food Technol.* 46:152.

Sime, W. J. (1990). Alginates. In *Food Gels*, Harris, P. (Ed.), pp. 53–78. Elsevier, Amsterdam/New York.

Sirine, G. (1968). Microencapsulation: a technique for limitless products. *Food Product Dev.* April-May:30–31, 34.

Sisson, W. A. (1943). X-ray examination. In *Cellulose and Cellulose Derivatives*, Ott, E. (Ed.), pp. 272–280. Interscience, New York.

Sjoberg, A.-M., and Pyysalo, H. (1985). Identification of food thickeners by monitoring of their pyrolytic products. *J. Chromatogr.* 319:90–98.

Smith, A. W., and Cooper, J. N. (1957). *The Elements of Physics*, 6th ed. McGraw-Hill, New York.

Smith, F., and Montgomery, R. (1956). End group analysis of polysaccharides. *Methods of Biochemical Analysis*, Vol. 3, pp. 153–212. Interscience, New York.

Smith, F., and Montgomery, R. (1959). *The Chemistry of Plant Gums and Mucilages*, pp. 77–132. Reinhold, New York.

Smith, N. O. (1982). *Elementary Statistical Thermodynamics*. Plenum, New York.

Snell, F. D., and Snell, C. T. (1953). *Colorimetric Methods of Analysis*, 3rd ed., Vol. III, p. 463. Van Nostrand, New York.

Sorochan, V. D., Dzizenko, A. K., Bodin, N. S., and Ovodov, Y. S. (1971). Light-scattering studies of pectic substances in aqueous solution. *Carbohydr. Res.* 20:243–249.

Sperling, L. H. (1986). *Introduction to Physical Polymer Science*. Wiley, New York.

Spiegel, J. E., Rose, R., Karabell, P., Frankos, V. H., and Schmitt, D. F. (1994). Safety and benefits of fructooligosaccharides as food ingredients. *Food Technol.* 48:85–89.

Sprenger, M. (1990). New stabilizing systems using galactomannans. *Dairy Ind. Int.* 55:20–21.

Srivastava, V. K., and Rai, R. S. (1963). Physico-chemical studies on gum dhawa (*Anogeissus latifolia* Wall). *Kolloid Z. Z. Polym.* 190:140–143.

Stacey, K. A. (1956). *Light Scattering in Physical Chemistry*, p. 176. Butterworth, London.

Stainsby, G. (1980). Proteinaceous gelling systems and their complexes with polysaccharides. *Food Chem.* 6:3–14.

Stanley, N. F. (1990). Carrageenans. In *Food Gels*, Harris, P. (Ed.), p. 94. Elsevier, Amsterdam/New York.

Stannett, V. T. (1989). Some recent developments in cellulose science and technology in North America. In *Cellulose*, Kennedy, J. F., Phillips, G. O., and Williams, P. A. (Eds.), pp. 20–31. Ellis Horwood, Chichester.

Stauffer, K. R. (1980). Gum tragacanth. *Handbook of Water-Soluble Gums and Resins*, Davidson, R. L. (Ed.), pp. 11-1–11-31. McGraw-Hill, New York.

Stauffer, K. R., Leeder, J. G., and Wang, S. S. (1980). Characterization of Zooglan-115, an exocellular glycan of Zoogloena ramigera-115. *J. Food Sci.* 45:946–952.

Steffe, J. F. (1992). *Rheological Methods in Food Process Engineering*. Freeman Press, East Lansing, MI.

Stelzer, G. I., and Klug, E. D. (1980). Carboxymethylcellulose. In *Handbook of Water-Soluble Gums and Resins*, Davidson, R. L. (Ed.), p. 4-4. McGraw-Hill, New York.

Stinson, E. E., and Willits, C. O. (1965). Isolation and characterization of the high molecular weight brown colorant of maple syrup. *J. Agric. Food Chem.* 13:294–297.

Stipanovic, A. J., and Giammatteo, P. J. (1989). Curdlan and scleroglucan. In *Polymers in Aqueous Media*, Glass, J. E. (Ed.), Advances in Chemistry 223, p. 73. Am. Chem. Soc., Washington, DC.

Stokke, B. T., and Elgsaeter, A. (1991). Electron microscopy of carbohydrate polymers. In *Advances in Carbohydrate Analysis*, White, C. A. (Ed.), Vol. 1, pp. 195–247. JAI Press, London.

Sun, S. F. (1994). *Physical Chemistry of Macromolecules*. Wiley, New York.

Szczesniak, A. S. (1983). Physical properties of foods: what they are and their relation to other food properties. In *Physical Properties of Foods*, Peleg, M., and Bagley, E. B. (Eds.), pp. 1–41. AVI Press, New York.

Szejtli, J. (1981). Cyclodextrins in foods, cosmetics and toiletries. In *Cyclodextrins*, Szejtli, J. (Ed.), pp. 469–480. Reidel, Boston.

Szejtli, J. (1991). *Biotechnology of Amylodextrin Oligosaccharides*, ACS Symposium Series 458. Am. Chem. Soc., Washington, DC.

Takahashi, S., and Seib, P. A. (1988). Paste and gel properties of prime corn and wheat starches with and without lipids. *Cereal Chem.* 65:474–483.

Tako, M., and Nakamura, S. (1984). Rheological properties of deacetylated xanthan in aqueous media. *Agric. Biol. Chem.* 48:2987–2993.

Tanaka, T. (1981). Gels. *Scientific American* 244:124–138.

Tanaka, T. (1992). Phase transitions of gels. In *Polyelectrolyte Gels*, Harland, R. S., and Prud'homme, R. K. (Eds.), ACS Symposium Series 480, pp. 1–21. Am. Chem. Soc., Washington, DC.

Tanford, C. (1961). *Physical Chemistry of Macromolecules*. Wiley, New York.

Tang, W. K., and Neill, W. K. (1964). Effect of flame retardants on pyrolysis and combustion of α-cellulose. *J. Polym. Sci. C* 6:65–81.

Teranishi, R., Takeoka, G. R., and Guntert, M. (1992). *Flavor Precursors: Thermal and Enzymatic Conversions,* ACS Symposium Series 490. Amer. Chem. Soc., Washington, DC.

Thayer, A. M. (1992). Food additives. *Chem. Eng. News* June:26–32, 35–38, 41–44.

Therkelsen, G. H. (1993). Carrageenan. In *Industrial Gums*, 3rd ed., Whistler, R. L., and BeMiller, J. N. (Eds.), pp. 145–180. New York, Academic Press.

Thibault, J.-F., Guillon, F., and Rombouts, F. M. (1991). Gelation of sugar beet pectin by oxidative coupling. In *The Chemistry and Technology of Pectin,* Walter, R. H. (Ed.), pp. 119–133. Academic Press, New York.

Thom, D., Dea, I. C. M., Morris, E. R., and Powell, D. A. (1982). Interchain associations of alginate and pectins. In *Gums and Stabilizers for the Food Industry,* Phillips, G. O., Wedlock, D. J., and Williams, P. A. (Eds.), Vol. 6, pp. 97–108. IRL Press, Oxford.

Thomas, J. W., Brown, D. L., Hoch, D. J., Leary, J. J., III, and Dokladalova, J. (1990). Determination of polydextrose (polymer) and residual monomers in polydextrose by liquid chromatography. *J. Assoc. Offic. Anal. Chem.* 74:571–573.

Thompson, A., and Wolfrom, M. L. (1958). The composition of pyrodextrins. *J. Am. Chem. Soc.* 80:6618–6620.

Thompson, D. B. (1992). Structure and functionality of carbohydrate hydrocolloids in food systems. In *Developments in Carbohydrate Chemistry,* Alexander, R. J., and Zobel, H. F. (Eds.), pp. 315–342. Am. Assoc. Cereal Chem., St. Paul, MN.

Toft, K. (1982). Interactions between pectins and alginates. In *Progress in Food and Nutritional Science,* Phillips, G. O., Wedlock, D. J., and Williams, P. A. (Eds.), Vol. 6, p. 95. Pergamon, Elmsford, NY.

Toft, K., Grasdalen, H., and Smidsrod, O. (1986). Synergistic gelation of alginates and pectins. In *Chemistry and Function of Pectins,* ACS Symposium Series 310, pp. 117–132. Am. Chem. Soc., Washington, DC.

Tolstoguzov, V. B. (1986). Functional properties of protein-polysaccharide mixtures. In *Functional Properties of Food Macromolecules,* Mitchell, J. R., and Ledward, D. A. (Eds.), pp. 385–415. Elsevier, Amsterdam/New York.

Tolstoguzov, V. B. (1993). Thermodynamic incompatibility of food macromolecules. In *Food Colloids and Polymers: Stability and Mechanical Properties,* Dickinson, E., and Walstra, P. (Eds.), pp. 94–102. Royal Chem. Soc., London.

Tomasik, P., Palasinski, M., and Wiejak, S. (1989a). The thermal decomposition of carbohydrates. Parts I. The decomposition of mono-, di-, and oligosaccharides. *Adv. Carbohydr. Chem. Biochem.* 47:203–278.

Tomasik, P., Wiejak, S., and Palasinski, M. (1989b). The thermal decomposition of carbohydrates. Part II. The decomposition of starch. *Adv. Carbohydr. Chem. Biochem.* 47:279–343.

Tomomatsu, H. (1994). Health effects of oligosaccharides. *Food Technol.* 48:61–65.

Topping, D. L. (1994). Physiological aspects of non-starch polysaccharides and resistant starch. International Food Hydrocolloid Conference, Ohio State University, Sept. 6–10, Columbus, OH.

Traitler, H., Del Vedovo, S., and Schweizer, T. F. (1984). Gas chromatographic separation of sugars by on-column injection on glass capillary columns. *J. High Res. Chromatogr. Chromatogr. Comm.* 7:558–562.

Trudso, J. E. (1989). Hydrocolloids—What can they do—How are they selected? *Can. Inst. Food Sci. Technol.* 21:AT/229-AT/235. Copenhagen Pectin A/S, DK-4623 Lille Skensved, Hercules, Inc., Denmark, 1992.

Turquois, T., Rochas, C., and Taravel, F. R. (1992). Rheological studies of synergistic kappa carrageenan-carob galactomannan gels. *Carbohydr. Polym.* 17:263–268.

Tvaroska, I. T., Rochas, C., Taravel, F. R., and Turquois, T. (1992). Kappa carrageenan-mannan interaction: a theoretical approach. In *Gums and Stabilizers for the Food Industry,* Phillips, G. O., Williams, P. A., and Wedlock, D. A. (Eds.), Vol. 6, pp. 231–234. IRL Press, Oxford.

Tye, R. J. (1988). Interactions between carrageenan and polyols and their application. In *Gums and Stabilizers for the Food Industry*, Phillips, G. O., Williams, P. A., and Wedlock, D. A. (Eds.), Vol. 4, pp. 293–300. IRL Press, Oxford.

Tye, R. J. (1991). Konjac flour: properties and applications. *Food Technol.* 45:82–92.

Ulrich, R. D. (1975). Membrane osmometry. In *Polymer Molecular Weights*. Part I, Slade, P. E., Jr. (Ed.), pp. 9–30. Dekker, New York.

Umano, K., and Shibamoto, T. (1984). Chemical studies on heated starch/glycine model systems. *Agric. Biol. Chem.* 48:1387–1393.

Van de Ven, T. G. M. (1989). *Colloidal Hydrodynamics*. Academic Press, New York.

van Fliet, T., Luyten, H., and Walstra, P. (1991). Fracture and yielding of gels. In *Food Polymers, Gels and Colloids*, Dickinson, E. (Ed.), Special Publication 82, pp. 392–403. Royal Chem. Soc., London.

van Oss, C. J. (1991). Interaction forces between biological and other polar entities in water: How many different primary forces are there? *J. Dispersion Sci. Technol.* 12:201–219.

Van Wazer, J. R., Lyons, J. W., Kim, K. Y., and Colwell, R. E. (1963). *Viscosity and Flow Measurement*. Interscience, New York.

Veis, A., and Eggenberger, D. N. (1954). Light scattering in solutions of a linear polyelectrolyte. *J. Am. Chem. Soc.* 76:1560–1563.

Vercellotti, J. R., and Crippen, K. L. (1991). Sugars as sources of flavors and aromas. Carbohydr. Div., Amer. Chem. Soc., Washington, DC, Abstract 66.

Viles, F. J., Jr., and Silverman, L. (1949). Determination of starch and cellulose with anthrone. *Anal. Chem.* 21:950–953.

Vink, H. (1954). Viscometry. In *High Polymers*, 2nd ed., Bikales, N. M., and Segal, L. (Eds.), Vol. V, Part IV, pp. 469–489. Wiley-Interscience, New York.

Vold, R. D., and Vold, M. J. (1983). *Colloid and Interface Chemistry*. Addison-Wesley, Reading, MA.

Voragen, A. G. J., Schols, H. A., and Pilnik, W. (1982). H. P. L. C. Analysis of anionic gums. *Prog. Food Nutr. Sci.* 6:379–385.

Wagner, R. H. (1949). Determination of osmotic pressure. In *Physical Methods of Organic Chemistry*, 2nd ed., Weissberger, A. (Ed.), Vol. 1, Part 1, pp. 494–499. Interscience, New York.

Walstra, P., van Fliet, T., and Bremer, L. G. B. (1991). On the fractal nature of particle gels. In *Food Polymers, Gels and Colloids*, Dickinson, E. (Ed.), pp. 369–382. Royal Chem. Soc., London.

Walter, R. H. (1991). Analytical and graphical methods for pectin. In *The Chemistry and Technology of Pectin*, Walter, R. H. (Ed.), pp. 189–225. Academic Press, New York.

Walter, R. H., and Fagerson, I. S. (1968). Volatile compounds from heated glucose. *J. Food Sci.* 33:294–297; erratum 33:654.

Walter, R. H., and Fagerson, I. S. (1970). Rates of pH change and disappearance of glucose during low temperature pyrolysis. *J. Food Sci.* 35:606–607.

Walter, R. H., and Jacon, S. A. (1994). Molecular weight approximations of ionic polysaccharides by pH determinations. *Food Hydrocoll.* 8:469–480.

Walter, R. H., and Matias, H. L. (1989). Volume fraction of a dispersed pectin. *Food Hydrocoll.* 3:205–208.

Walter, R. H., and Seeger, S. C. (1990). Water activity and moisture content of selected foods of commerce in Hawaii. *J. Food Protection* 53:72–74.

Walter, R. H., and Sherman, R. M. (1981). Apparent activation energy of viscous flow in pectin jellies. *J. Food Sci.* 46:1223–1225.

Walter, R. H., and Sherman, R. M. (1983). The induced stabilization of aqueous pectin dispersions by ethanol. *J. Food Sci.* 48:1235–1237, 1241.

Walter, R. H., and Sherman, R. M. (1986). Rheology of high-methoxyl pectin jelly sols prepared above and below the gelation temperature. *Lebensm. Wiss. Technol.* 19:95–100.

Walter, R. H., and Sherman, R. M. (1988). Application of intrinsic viscosity and the interaction coefficient to some ionic galacturonan dispersions. *Food Hydrocoll.* 2:151–158.

Walter, R. H., and Talomie, T. G. (1990). Quantitative definition of polysaccharide hydrophilicity. *Food Hydrocoll.* 4:197–203.

Walter, R. H., Rao, M. A., Sherman, R. M., and Cooley, H. J. (1985). Edible fibers from apple pomace. *J. Food Sci.* 50:747–749.

Walter, R. H., Rao, M. A., Van Buren, J. P., Sherman, R. M., and Kenney, J. F. (1977). Development and characterization of an apple cellulose gel. *J. Food Sci.* 42:241–243.

Walter, R. H., Van Buren, J. P., and Sherman, R. M. (1978). Dispersion of pectin in dimethyl sulfoxide. *J. Food Sci.* 43:1882–1883.

Ward-Smith, R. S., Hey, M. J., and Mitchell, J. R. (1994). Protein–polysaccharide interactions at the oil–water interface. *Food Hydrocoll.* 8:309–315.

Watase, M., and Nishinari, K. (1987). Rheological and thermal properties of carrageenan gels. *Makromol. Chem.* 188:2213–2221.

Watson, S. A. (1964). Corn starch. *Methods in Carbohydrate Chemistry*, Vol. IV, pp. 3–5. Academic, New York.

Wedlock, D. J., Fasihuddin, B. A., and Phillips, G. O. (1986). Comparison of molecular weight determination of sodium alginate by sedimentation–diffusion and light scattering. *Int. J. Biol. Macromol.* 8:57–61.

Weibel, M. K. (1994). Microdisassembled cellulose as a new food ingredient. Internat. Food Hydrocoll. Conf., Ohio State Univ., Sept. 6–10, Columbus, OH, Abstract.

West, E. S., and Todd, W. R. (1961). *Textbook of Biochemistry*, 3rd ed., p. 26. Macmillan, New York.

Wheaton, R. M., and Bauman, W. C. (1953). Ion exclusion. A unit operation utilizing ion exchange materials. *Ind. Eng. Chem.* 45:228–233.

Whelan, W. J. (1964). Determination of reducing end-groups. *Methods in Carbohydrate Chemistry*, Vol. IV, pp. 72–78. Academic Press, New York.

Whistler, R. L. (1993). Exudate gums. In *Industrial Polysaccharides*, Whistler, R. L., and BeMiller, J. N. (Eds.), p. 313. Academic Press, New York.

Whistler, R. L., and Feather, M. S. (1965). Hemicellulose extraction from annual plants with alkaline solutions. *Methods in Carbohydrate Chemistry*, Vol. V, pp. 144–145. Academic Press, New York.

Whistler, R. L., and Smart, C. L. (1953). *Polysaccharide Chemistry*. Academic Press, New York.

White, A. R. (1982). Visualization of cellulases and cellulose degradation. In *Cellulose and Other Natural Polymer Systems*, Brown, R. M., Jr. (Ed.), pp. 489–509. Plenum, New York.

Williams, P. A., Day, D. H., Langdon, M. J., Phillips, G. O., and Nishinari, K. (1991). Synergistic interaction of xanthan gum with glucomannans and galactomannans. *Food Hydrocoll.* 4:489–493.

Williams, V. R., Mattice, W. L., and Williams, H. B. (1978). *Basic Physical Chemistry for the Life Sciences*, 3rd ed. Freeman, San Francisco.

Williamson, S., and McCormick, C. L. (1994). Derivatization of chitin and cellulose utilizing LiCl/N, *N*-dimethylacetamide. 208th Annual Meeting, Am. Chem. Soc., Div. Cellulose, Paper and Textile, August 21–26, Washington, DC.

Wilson, R. H., Goodfellow, B. J., and Belton, P. S. (1988). Fourier transform infrared spectroscopy and biopolymer functionality. In *Gums and Stabilizers for the Food Industry*, Phillips, G. O., Williams, P. A., and Wedlock, D. A. (Eds.), Vol. 4, pp. 81–87. IRL Press, Oxford.

Wolf, M. J. (1964). Wheat starch. *Methods in Carbohydrates Chemistry*, Vol. IV, pp. 6–9. Academic Press, New York.

Wong, D. W. S. (1989). *Mechanism and Theory in Food Chemistry*, p. 106. Van Nostrand–Reinhold, New York.

Wong, K. K. Y., and Saddler, J. N. (1993). Application of hemicellulases in the food, feed, and pulp and paper industries. In *Hemicellulose and Hemicellulases*, Coughlan, M. P., and Hazlewood, G. P. (Eds.), pp. 127–143. Res. Monograph, Portland Press, Chapel Hill, NC.

Wunderlich, B. (1990). *Thermal Analysis*. Academic Press, New York.

Wurzburg, O. B. (Ed.) (1986). *Modified Starches: Properties and Uses*. CRC Press, Boca Raton, FL.

Wurzburg, O. B. (1995). Modified starches. In *Food Polysaccharides and Their Applications*, Stephen, A. M. (Ed.), pp. 67–97. Dekker, New York.

Yackel, W. C., and Cox, C. (1992). Application of starch-based fat replacers. *Food Technol.* 46:146.

Yakubu, P. I., Baianu, I. C., and Orr, P. H. (1990). Unique hydration behavior of potato starch as determined by deuterium nuclear magnetic resonance. *J. Food Sci.* 55:458–461.

Yalpani, M. (1988). Polysaccharides, pp. 405–479. Elsevier, Amsterdam/New York.

Young, S., and Torres, J. A. (1989). Xanthan: effect of molecular conformation on surface tension properties. *Food Hydrocoll.* 3:365–377.

Zhao, J., and Whistler, R. L. (1994). Spherical aggregates of starch granules as flavor carriers. *Food Technol.* 48:104ff.

Zimm, B. H. (1948). The scattering of light and the radial distribution function of high polymer solutions. *J. Chem. Phys.* 16:1093–1099; Apparatus and methods for measurement and interpretation of the angular variation of light scattering; preliminary results on polystyrene solutions. *J. Chem. Phys.* 16:1099–1116.

Zimm, B. H. (1956). Dynamics of molecules in dilute solution: viscoelasticity, flow birefringence and dielectric loss. *J. Chem. Phys.* 24:269–278.

Zimm, B. H., and Stockmayer, W. H. (1949). The dimensions of chain molecules containing branches and rings. *J. Chem. Phys.* 17:1301–1314.

Zuber, M. S. (1965). Genic control of starch development. In *Starch: Chemistry and Technology*, Whistler, R. L., and Paschall, E. F. (Eds.), pp. 43–63. Academic Press, New York.

Index

A

D

E

F

G

H

I

J

K

Q

R

S

T

FOOD SCIENCE AND TECHNOLOGY

International Series

Maynard A. Amerine, Rose Marie Pangborn, and Edward B. Roessler, *Principles of Sensory Evaluation of Food*. 1965.

Martin Glicksman, *Gum Technology in the Food Industry*. 1970.

Maynard A. Joslyn, *Methods in Food Analysis*, second edition. 1970.

C. R. Stumbo, *Thermobacteriology in Food Processing*, second edition. 1973.

Aaron M. Altschul (ed.), *New Protein Foods*: Volume 1, *Technology*, *Part A*—1974. Volume 2, *Technology*, *Part B*—1976. Volume 3, *Animal Protein Supplies*, *Part A*—1978. Volume 4, *Animal Protein Supplies*, *Part B*—1981. Volume 5, *Seed Storage Proteins*—1985.

S. A. Goldblith, L. Rey, and W. W. Rothmayr, *Freeze Drying and Advanced Food Technology*. 1975.

R. B. Duckworth (ed.), *Water Relations of Food*. 1975.

John A. Troller and J. H. B. Christian, *Water Activity and Food*. 1978.

A. E. Bender, *Food Processing and Nutrition*. 1978.

D. R. Osborne and P. Voogt, *The Analysis of Nutrients in Foods*. 1978.

Marcel Loncin and R. L. Merson, *Food Engineering*: *Principles and Selected Applications*. 1979.

J. G. Vaughan (ed.), *Food Microscopy*. 1979.

J. R. A. Pollock (ed.), *Brewing Science*, Volume 1—1979. Volume 2—1980. Volume 3—1987.

J. Christopher Bauernfeind (ed.), *Carotenoids as Colorants and Vitamin A Precursors*: *Technological and Nutritional Applications*. 1981.

Pericles Markakis (ed.), *Anthocyanins as Food Colors*. 1982.

George F. Stewart and Maynard A. Amerine (eds.), *Introduction to Food Science and Technology*, second edition. 1982.

Malcolm C. Bourne, *Food Texture and Viscosity*: *Concept and Measurement*. 1982.

Hector A. Iglesias and Jorge Chirife, *Handbook of Food Isotherms*: *Water Sorption Parameters for Food and Food Components*. 1982.

Colin Dennis (ed.), *Post-Harvest Pathology of Fruits and Vegetables*. 1983.

P. J. Barnes (ed.), *Lipids in Cereal Technology*. 1983.

David Pimentel and Carl W. Hall (eds.), *Food and Energy Resources*. 1984.

Joe M. Regenstein and Carrie E. Regenstein, *Food Protein Chemistry*: *An Introduction for Food Scientists*. 1984.

Maximo C. Gacula, Jr., and Jagbir Singh, *Statistical Methods in Food and Consumer Research*. 1984.

Fergus M. Clydesdale and Kathryn L. Wiemer (eds.), *Iron Fortification of Foods*. 1985.

Robert V. Decareau, *Microwaves in the Food Processing Industry*. 1985.

S. M. Herschdoerfer (ed.), *Quality Control in the Food Industry*, second edition. Volume 1—1985. Volume 2—1985. Volume 3—1986. Volume 4—1987.

F. E. Cunningham and N. A. Cox (eds.), *Microbiology of Poultry Meat Products*. 1987.

Walter M. Urbain, *Food Irradiation*. 1986.

Peter J. Bechtel, *Muscle as Food*. 1986.

H. W.-S. Chan, *Autoxidation of Unsaturated Lipids*. 1986.

Chester O. McCorkle, Jr., *Economics of Food Processing in the United States*. 1987.

Jethro Jagtiani, Harvey T. Chan, Jr., and William S. Sakai, *Tropical Fruit Processing*. 1987.

J. Solms, D. A. Booth, R. M. Dangborn, and O. Raunhardt, *Food Acceptance and Nutrition*. 1987.

R. Macrae, *HPLC in Food Analysis*, second edition. 1988.

A. M. Pearson and R. B. Young, *Muscle and Meat Biochemistry*. 1989.

Dean O. Cliver (ed.), *Foodborne Diseases*. 1990.

Marjorie P. Penfield and Ada Marie Campbell, *Experimental Food Science*, third edition. 1990.

Leroy C. Blankenship, *Colonization Control of Human Bacterial Enteropathogens in Poultry*. 1991.

Yeshajahu Pomeranz, *Functional Properties of Food Components*, second edition. 1991.

Reginald H. Walter, *The Chemistry and Technology of Pectin*. 1991.

Herbert Stone and Joel L. Sidel, *Sensory Evaluation Practices*, second edition. 1993.

Robert L. Shewfelt and Stanley E. Prussia, *Postharvest Handling: A Systems Approach*. 1993.

R. Paul Singh and Dennis R. Heldman, *Introduction to Food Engineering*, second edition. 1993.

Tilak Nagodawithana and Gerald Reed, *Enzymes in Food Processing*, third edition. 1993.

Dallas G. Hoover and Larry R. Steenson, *Bacteriocins*. 1993.

Takayaki Shibamoto and Leonard Bjeldanes, *Introduction to Food Toxicology*. 1993.

John A. Troller, *Sanitation in Food Processing*, second edition. 1993.

Ronald S. Jackson, *Wine Science: Principles and Applications*. 1994.

Harold D. Hafs and Robert G. Zimbelman, *Low-fat Meats*. 1994.

Lance G. Phillips, Dana M. Whitehead, and John Kinsella, *Structure–Function Properties of Food Proteins*. 1994.

Robert G. Jensen, *Handbook of Milk Composition*. 1995.

Yrjö H. Roos, *Phase Transitions in Foods*. 1995.